ANNALS *of* THE NEW YORK ACADEMY OF SCIENCES

T0188409

EDITOR-IN-CHIEF
Douglas Braaten

ASSOCIATE EDITOR
Rebecca E. Cooney

PROJECT MANAGER
Steven E. Bohall

Artwork and design by Ash Ayman Shairzay

The New York Academy of Sciences
7 World Trade Center
250 Greenwich Street, 40th Floor
New York, NY 10007-2157

annals@nyas.org
www.nyas.org/annals

The New York
Academy of Sciences

Published by Blackwell Publishing
On behalf of the New York Academy of Sciences

Boston, Massachusetts
2012

ANNALS *of* THE NEW YORK ACADEMY OF SCIENCES

VOLUME
1272

ISSUE

Advances Against Aspergillosis I

Clinical Science

ISSUE EDITORS

Karl V. Clemons,[a] Malcolm Richardson,[b] and David S. Perlin[c]

[a]California Institute for Medical Research and Stanford University, [b]University of Manchester, and [c]UMDNJ–New Jersey Medical School

TABLE OF CONTENTS

Ann. N.Y. Acad. Sci. ISSN 0077-8923

ANNALS OF THE NEW YORK ACADEMY OF SCIENCES
Issue: *Advances Against Aspergillosis*

Preface for *Advances Against Aspergillosis*

The fungal genus *Aspergillus* comprises hundreds of species and is an extremely complex taxonomic collection of organisms, including some cryptic species. These molds are ubiquitous in nature, found on decaying organic matter, in the soil, and in marine environments. The saprophytic aspect of *Aspergillus* is very important with respect to environmental ecology. In relationship to humans, *Aspergillus* is important for agricultural, industrial, and medical reasons. In agricultural settings, contamination of food crops with aflatoxin, produced by toxigenic *Aspergillus flavus* is a worldwide problem, which results in the subsequent increased exposure of humans consuming these foods to aflatoxin (a carcinogen). An industrial application of certain *Aspergillus* species is their use in performing specific fermentations, such as in the manufacturing of sake. Medically, about 20 species of *Aspergillus* have been shown to cause disease, with a single species, *A. fumigatus*, responsible for the vast majority of infections and allergies. Although usually an opportunistic pathogen in humans, *A. fumigatus* can be a primary pathogen in animals. Invasive pulmonary aspergillosis is particularly important as a cause of a rapidly fatal disease in avian populations like chickens and turkeys. For humans, aspergillosis has become the fourth leading opportunistic infection of immunocompromised individuals. It is a primarily invasive pulmonary disease, but about one-third of patients have multiple organ involvement, with CNS disease being frequent and often fatal. In spite of advances in antifungal therapy, morbidity and mortality remain high. In addition, *Aspergillus* species cause other diseases in humans that include keratitis, allergic bronchopulmonary aspergillosis in asthma or cystic fibrosis, otomycosis, and several forms of rhinosinusitis and mycotoxicosis.

Recognizing the growing importance of *Aspergillus*, particularly as a cause of human disease, Drs. David W. Denning, William J. Steinbach, and David A. Stevens organized and chaired the inaugural "Advances Against Aspergillosis" conference held in San Francisco in September 2004. With the success of this first meeting, they organized subsequent conferences that have been held every two years. The primary goals have been to bring together clinicians, nurses, clinical laboratorians, epidemiologists, geneticists, molecular biologists, and basic researchers, along with the leading experts in each of these areas from around the world, as well as the introduction of new topics and presenters, and the establishment of new areas of collaborative research and scientific interaction among the different scientific disciplines represented.

The 5th "Advances Against Aspergillosis" was held in Istanbul, Turkey on January 26–28, 2012, with 375 registered attendees from 39 countries and various educational backgrounds attending. The chairs and the scientific committee developed a program that included 40 invited talks, 6 chosen from submitted abstracts, and 144 posters. Sessions included management and treatment of aspergillosis, pathogenesis and immunology, new disease associations, basic biology, non-*fumigatus* *Aspergillus* species, genomics and proteomics, diagnostics, and novel antifungal agents. Poster presentations encompassed all aspects of the study of *Aspergillus*. The full program and programs from past conferences can be viewed at http://www.advancesagainstaspergillosis.org; additional slides and abstracts can be seen at http://www.aspergillus.org.uk.

The papers included in two volumes of *Annals of the New York Academy of Sciences* provide a cross-section of the information presented at the 5th "Advances Against Aspergillosis." Together with the chairs and the scientific committee, we wish to thank the eleven sponsors whose donations allowed the conference to be held, and for scholarships that were given to 28 deserving applicants; the speakers, poster presenters, and attendees, all of whom made the conference a success; the authors of the papers in the two volumes for their time and effort in writing reviews; and the individuals who gave freely of their time in reviewing the manuscripts. We also thank Dr. Douglas Braaten, editor-in-chief, *Annals*, for agreeing to publish the collection of reviews, and for the assistance that he and his staff provided. We look forward to convening the 6th "Advances Against Aspergillosis" in 2014; up-to-date information will be posted at http://www.advancesagainstaspergillosis.org.

KARL V. CLEMONS
Stanford University, Stanford, California

DAVID S. PERLIN
UMDNJ–New Jersey Medical School, Newark, New Jersey

MALCOLM RICHARDSON
University of Manchester, Manchester, United Kingdom

Ann. N.Y. Acad. Sci. ISSN 0077-8923

ANNALS OF THE NEW YORK ACADEMY OF SCIENCES
Issue: *Advances Against Aspergillosis*

Application of diagnostic markers to invasive aspergillosis in children

Emmanuel Roilides and Zoi-Dorothea Pana

Third Pediatric Department, Aristotle University School of Medicine, Hippokration General Hospital, Thessaloniki, Greece

Address for correspondence: Emmanuel Roilides, Third Department of Pediatrics, Hippokration General Hospital, Konstantinoupoleos 49, 54642, Thessaloniki, Greece. roilides@med.auth.gr

Early mycological detection of *Aspergillus* species is the cornerstone for a prompt diagnosis, appropriate treatment strategies, and improved survival of patients with invasive aspergillosis (IA), irrespective of age. However, the currently available laboratory tests for the diagnosis of IA include culture with direct microscopy, histology, and antigenic markers, such as galactomannan and β-1, 3-D-glucan, *Aspergillus* spp. DNA detection by PCR, and imaging studies, such as high-resolution CT scan. However, all need further validation, especially in children. In this review we focus on the diagnosis of IA emphasizing the current perspectives, difficulties in interpretation, and the need of further evaluation from a pediatric point of view.

Keywords: invasive aspergillosis; β-1, 3-D-glucan; galactomannan; children

Introduction

Invasive aspergillosis (IA) is an increasing problem in children and is associated with high attributable morbidity and mortality rates, as well as early and late onset complications.[1] There are significant differences reported in the epidemiology patterns of IA among different pediatric populations and adults.[2] Children with either congenital or acquired immunodeficiency may suffer from IA. Previous studies have emphasized the increased risk for IA in children with acquired immunodeficiencies after immunosuppressive therapy for cancer (especially for hematological malignancies), in children with advanced human immunodeficiency virus infection, bone marrow failure syndromes or allogeneic hematopoietic stem cell or solid organ transplantation.[3–5] Furthermore, children with inherited defects of phagocytic host defenses, including chronic granulomatous disease and those with cystic fibrosis are at increased risk of presenting IA.[6,7] Additionally, low-birth weight and premature infants are susceptible to IA due to reduced chemotactic, phagocytic, and microbicidal activity, skin breakdown, prior antibiotic therapy, or prolonged use of steroids.[8]

Early mycological detection of *Aspergillus* species is the cornerstone for a prompt diagnosis, appropriate treatment strategy, and improved survival of patients with IA, irrespective of age. On the other hand, current laboratory examinations for IA detection need further validation, especially in children. Routine methods for rapid specific identification of *Aspergillus* species are generally not available. The current diagnostic markers for IA include conventional and more recent methods under evaluation. Conventional methods of diagnosis include direct microscopy and histology, and culture of respiratory and various fluids and tissues. Recently, more rapid and sensitive methods have been developed, for example, the detection of antigenic markers, such as galactomannan and β-1, 3-D-glucan, the detection of molecular markers of *Aspergillus* DNA by polymerase chain reaction, and improved imaging studies such as high-resolution CT scans.

Various problems characterize the older and the more recently introduced diagnostic markers, especially in pediatric patients. A few examples of these problems include *in vitro* culture lacks sensitivity, histological diagnosis requires invasive methods that are often difficult in children and are nonspecific to speciation, galactomannan lacks sensitivity in children compared with adults,

doi: 10.1111/j.1749-6632.2012.06828.x

imaging lacks specificity, and fungal DNA detection requires standardization, especially for samples from children.[9] In this short review we focus on the diagnostic panel used for the diagnosis of IA, emphasizing the current perspectives, interpretation difficulties, and the need of further evaluation from a pediatric point of view. Also, we briefly mention the standard methods in which no differences in performance exist between adults and children.

Conventional methods

Conventional methods include direct microscopy and histology and culture of respiratory and other fluids and tissue. However, culture is insensitive and is time consuming.[10] The high percentage of culture-negative results may be attributed to the atypical appearances of *Aspergillus* spp. or to the lack of expertise for species identification. Additionally, the time required to obtain culture results delays the onset of antifungal treatment, leading to deterioration of survival outcomes. Tissue diagnosis also requires invasive procedures that are complicated with serious side effects, especially in patients with thrombocytopenia.

Molecular markers

The polymerase chain reaction (PCR) represents one of the most investigated rapid diagnostic methods with clinical utility for IA. A recent meta-analysis of the use of PCR from blood, serum, or plasma samples for the detection of IA reported that the sensitivity and specificity for two consecutive positive samples were 0.75 (95% CI 0.54–0.88) and 0.87 (95% CI 0.78–0.93), respectively.[11] A similar meta-analysis evaluating PCR on bronchoalveolar lavage fluid revealed a sensitivity of 0.91 (95% confidence interval (CI), 0.79–0.96) and specificity of 0.92 (95% CI, 0.87–0.96).[12] However, PCR has a number of drawbacks that include the lack of standardization and uniformity of methods and results obtained among laboratories. Moreover, the difficulty of interpretation of the findings obtained may result in inconsistencies and unreliable results in both adult and pediatric populations. To overcome these problems and to optimize PCR findings, a real-time quantitative PCR (qPCR) assay based on minimum information for the publication of real-time quantitative PCR experiments (MIQE) guidelines has been recently proposed to detect *Aspergillus* spp., with promising results.[13]

Other issues, such as appropriate sample timing, the frequency of prospective consecutive sampling, and the number of positive PCR results required to initiate antifungal treatment remain to be established.[14] A useful practice to improve the sensitivity of the assay is to obtain the samples before the introduction of antifungal therapy. A recent report on an *in vivo* model revealed that the sensitivity of both qPCR and GM in the early diagnosis phase of IA was significantly impacted by the use of posaconazole and caspofungin.[15] Furthermore, the definition of an episode of IA as PCR positive, with two positive results within 14 days in patients with hematopoietic stem cell transplantation and acute leukaemia, raised the sensitivity, specificity, positive predictive value, and negative predictive value of IA (100%, 75.4%, 46.4%, and 100%, respectively).[16]

As far as pediatric patients are concerned, data concerning DNA detection of *Aspergillus* spp. with different PCR techniques are lacking. Only one recent multicenter study conducted by Hummel *et al.* has evaluated with nested PCR assay 291 clinical samples from 71 pediatric and adolescent patients with suspected IA during the period 2000–2007.[17] Samples were obtained mainly from blood but cerebrospinal fluid and bronchoalveolar lavage samples were also included. According to their results, the sensitivity and specificity were high, with rates of 80% and 81%, respectively. Additionally, the reported positive and negative predictive values were 40% and 96%, respectively. A large multicenter study that is currently on-going to investigate the role of PCR in IA does not include children.

Novel molecular markers

The proteomic signature of specific *Aspergillus* spp. is thought to be paramount to widen future diagnostic panels and to develop more personalized therapeutic regimens for IA. Recent immunoproteomic studies have focused on the detection of new more accurate, novel immunodiagnostic markers of *Aspergillus* spp.[18–20] Several extracellular proteins of *Aspergillus* are highly immunogenic and may serve as specific diagnostic antigens in humans in the future. Among immunoactive proteins of *Aspergillus* spp., efforts have been undertaken to identify the one with the best immunogenic potential. A recent study by Shi *et al.*, based on immunoproteomic data, reported that 17 different extracellular proteins of

A. fumigatus represent novel candidate biomarkers for the future detection of IA.[18] In particular, thioredoxin reductase GliT (TR) presented the best immunogenic potential for *A. fumigatus*, with analysis showing low homology to other *Aspergillus* spp. and no homology to any human proteins.[19] In another study, a panel of three novel proteins could serve as potential allergens with specific IgE immunoreactivity in patients suffering from allergic bronchopulmonary aspergillosis (ABPA).[20] Essential virulent factors, such as siderophores, represent candidate markers for *Aspergillus* detection and simultaneously for developing therapeutic targeted inhibitors as novel antifungal agents.[21]

Galactomannan assay

Serum

The galactomannan assay (GM) appears to be an attractive serological diagnostic marker for detecting IA. Recent data have demonstrated that GM seems to be predictive of outcome in hematological patients with invasive pulmonary aspergillosis. In particular, high baseline serum GM antigen levels on day 45 posttransplantation were significantly associated with poor outcome.[22]

Among the GM tests developed, the double-sandwich ELISA is found to be the most accurate, irrespective of patient age. To date, no formal recommendations have been published for GM testing in serum specifically in the pediatric population. For that reason cut-off values are mainly extrapolated from adult populations. In addition, false-positivity of the GM test in children might be attributed to several factors, such as concomitant administration of various antibiotics (i.e., piperacillin/tazobactam), cross-reactivity with environmental moulds such as *Penicillium marneffei* or *Cryptococcus neoformans*, milk-based diet, and the presence of *Bifidobacterium bifidum*.[23]

The results of previously published studies have to be interpreted with caution since they suffer from heterogeneity of cut-off values, of definitions of assay positivity, and of the analyses performed (e.g., analyzing patients, episodes or some single sample results). Among 10 studies conducted (7 prospective) evaluating serum GM in children, the number of pediatric patients included in each study varied between 20 and 347, and the number of samples from 413 to 2376 (Table 1).[24–34] Furthermore, the number of patients with proven/probable invasive

fungal diseases (IFD), and of controls, also varied widely (median 9.5 (range, 1–28) and 63 (range 8–338), respectively). True positive GM results are additionally variable, ranging from 0 to 100% (4 studies with ≥10 patients with proven/probable IFD: 28–92% (median, 71.5%)), while true negative results fluctuated from 22 to 100% (7 studies with ≥10 controls: 49–100% (median, 88.5%)).

It is also interesting to note that in most studies serial GM testing was assessed in children with hematological malignancies and after allogeneic HSCT (screening performed once or twice weekly); the results show a sensitivity and specificity profile of GM testing that is similar to that observed in adults. However, in patients with inherited immunodeficiencies (i.e., chronic granulomatous disease) or other nonhematologic categories, there is usually lack of angioinvasion and presence of dysfunctional neutrophils and monocytes. In these cases, serum GM is often negative.[35,36]

The comparison of five studies, which use European Organization for Research and Treatment of Cancer (EORTC)/Mycosis Study Group (MSG) criteria and give adequate information for individual patients, with results of a formal meta-analysis of adult data conducted by Pfeiffer *et al.*, revealed that GM sensitivity in children is 0.76 (95% CI 0.62–0.87) compared to 0.73 (95% CI 0.46–0.61) in adults, while GM specificity reached 0.86 (95% CI 0.68–0.95) in children compared to 0.90 (95% CI 0.88–0.92) in adults, respectively.[37] In 2011 these studies were comprehensively analyzed and discussed by the European Conference on Infections in Leukemia (ECIL) 4 Group and its recommendations are reported electronically.[38] According to published data, prospective monitoring of GM every 3–4 days in children at high risk for IFD is reasonable for early diagnosis of IA, despite the number of limitations in the available pediatric data mentioned above (i.e., wide variations among the studies regarding cut-off, definition of positivity, etc.).[38,39] In addition, although the optimal cut-off value of GM in the serum of children is not well defined, ECIL 4 recommendations support the use of a threshold of an optical density index 0.5.[38]

Bronchoalveolar lavage and cerebrospinal fluid

The presence of galactomannan in bronchoalveolar lavage (BAL) fluid (BAL GM) is an alternative

Table 1. Serum galactomannan (GM) assay in studies with pediatric patients

Author Ref. no.	No. of ped. pts.	No. of GM samples	Definition of GM positivity	Type of collection	Cut-off values (ng/mL)	IFD definition	Results
Rohrlich et al.[24]	37	413	Two consecutive samples	Screening 2 × wk: during immunosuppression	≥0.93	Guiot CID 1994	PPV: 83%, FP: 5.7% Clinical + radiol. signs occurred at a mean of 13.4 days after GM (+)
Sulahian et al.[25]	347	2376	Two consecutive samples	Screening 2 × wk: during immunosuppression	≥1.5	EORTC/MSG	SP: 89.9% SN: 10% FP: 10.1%
Herbrecht et al.[26]	NR**	540	Per sample	Screening 2 × wk: during immunosuppression	≥1.5	EORTC/MSG	SP: 47.6% PPV: 15.4% NPV: 96.7% FP results more frequently in Chil. [44.0%] than in adults [0.9%]; $P = 0.0001$. Most FP occurred in HSCT pts.
Challier et al.[27]	20	207	NR	NR	≥1	EORTC/MSG	SN and SP about 90% with clinical and radiological symptoms and 50% in absence
El Mahallawy et al.[28]	91	NR	NR	NR	NR	EORTC/MSG	SN: 79% SP: 61% PPV: 54% NPV: 83%
Hovi et al.[29]	98	932	NR	Screening 1 × wk: during immunosuppression	NR	EORTC/MSG	NR
Steinbach et al.[30]	64	826	Per sample	Screening 2 × wk: during immunosuppression	≥0.5	EORTC/MSG	SP: 97.5%§ Exclusion of pts. receiving PIP/ TAZ = SP: 98.4%
Hayden et al.[31]	56	990	Per sample	Screening 1 × wk: during immunosuppression	≥0.5	EORTC/MSG	SN: 65.7% GM may precede clin. + microbial. + radiolog. evidence of IA.
Armenian et al.[32]	68	1086	Two consecutive samples	Screening 2 × wk: during immunosuppression	≥0.5	EORTC/MSG	13 samples (1.2%) from 4 patients (5%) were GM+.
Castagnola et al.[33]	119	1798	Per sample or two consecutive samples	Not specified At least 2 × week	≥0.7 for single test or 0.5-0.72 for consecutive tests	EORTC/MSG	SP: 98%, SN: 32% PPV: 70%, NPV: 92% Better GM performance after chemother. than after HSCT
de Mol et al.[34]	41	41	Per sample		≥0.5	EORTC/MSG	Among 13 pts with a pos. serum GM, 11 had a pos. BAL GM (Spearman's Coeff. 0.719). BAL GM SN: 82.4% SP: 87.5%. PPV: 82.4% NPV: 87.5%

*NR: not reported, § SN: sensitivity, SP: specificity, PPV: positive predictive value, NPV: negative predictive value, FP: false positive.
**48 episodes in children.

serological diagnostic marker, especially for invasive pulmonary aspergillosis, which constitutes the most common presentation of IA. The reported sensitivity rate of BAL GM is in general higher than that in serum due to the increased fungal burden in the bronchi of patients with pulmonary IA.[39] However, the role of BAL GM in pediatric IA has not been extensively evaluated. A recent retrospective study conducted by Desai et al. suggests that BAL GM is a valuable adjunctive diagnostic tool.[40] According to their results from 85 pediatric patients, 59 of whom were immunocompromised (9 with proven/probable invasive pulmonary aspergillosis), the receiver operating characteristic curves for the immunocompromised patients demonstrated an optimal BAL GM OD cut-off value of 0.87, that yielded a sensitivity for probable/proven IA of 78% and a specificity of 100%. At 0.87, the positive and negative predictive values among immunocompromised patients were 58% and 96%, respectively. According to the recommendations of the 2011 ECIL 4 Group, the limited published data support the value of GM in the diagnosis of pulmonary aspergillosis (BAL GM cut-off value 1) in children.[38] Nevertheless, systemic mold-active antifungal prophylaxis may decrease the performance of the test.[38]

Although invasive pulmonary aspergillosis is the most frequent presentation of IA, cerebral aspergillosis also may occur, especially in severely immunocompromised patients undergoing HSC transplantation. GM can be found in the cerebrospinal fluid (CSF) but the data are very limited and based on only a few case reports or case series. A recent case series study by Viscoli et al. evaluated GM in CSF samples of five patients with probable cerebral aspergillosis and of 16 control patients.[41] According to their results the median CSF GM index for the five patients with probable cerebral aspergillosis (eight samples) was 10.52 (mean, 81.93; standard deviation (SD), 200.96; range, 0.55 to 578.50). In contrast, the median CSF GM index from the 16 control patients (33 samples) was 0.29 (mean, 0.28; SD, 0.13; range, 0.09 to 0.75). A previous case report described a rare pediatric case of an 18-month infant suffering from corticosteroid-resistant nephrotic syndrome, which was complicated by invasive pulmonary and cerebral aspergillosis.[42] The reported GM value was 4 ng/mL in CSF and 3.5 ng/mL in serum (normal value <1 ng/mL). Although further studies on larger patient populations and especially in children are needed, the recommendations of the ECIL 4 Group support the value of GM in the diagnosis of central nervous system aspergillosis (CSF GM cut-off 0.5).[38]

Beta-D-glucan assay

The β-1, 3-D-glucan assay (BDG) is also thought to have clinical utility, both for surveillance and as a diagnostic tool for IA. The BDG is a nonspecific diagnostic marker that can be detected by two commercial assays (Fungitell®, Associates of Cape Cod; Fungitec G®, Seikagaku) in fungal infections attributed to Aspergillus and Candida spp., as well as those due to Fusarium, Trichosporum, Saccharomyces, and Pneumocystis jirovecii. Furthermore, bacteria, such as Streptococcus pneumonia and Pseudomonas aeruginosa, and sometimes healthy individuals, might present positive BDG test results.[43] On the other hand, BDG is absent in cryptococcosis and mucormycosis. False-positive BDG has been associated with the concomitant use of antibiotics, such as cefepime, piperacillin/tazobactam, or meropenem.[43]

According to the revised EORTC/MSG definitions, BDG is included as a mycological criterion of invasive fungal disease. In adult populations especially, the BDG test presents a satisfactory diagnostic for early IFD.[43–45] Based on data in 2979 adult patients (594 with proven or probable IFD) the pooled sensitivity of BDG reached 76.8% (95% CI 67.1%–84.3%) and the pooled specificity of 85.3% (95% CI, 79.6%–89.7%), respectively. However, data regarding pediatric populations for IA are extremely limited. In a recent study conducted by Mularoni et al. the BDG test was evaluated in four children with proven IFD (three patients with candidemia, one patient with probable aspergillosis) by positive culture from a sterile site and/or the demonstration of fungal elements in diseased tissues.[46] The BDG test was performed using the Fungitell assay with a positive cut-off of 60 pg/mL. For all four patients BDG levels were >523 pg/mL. A major problem in performing a BDG test on samples from children is the lack of baseline levels for uninfected pediatric patients. For that reason, Smith et al. tried to evaluate the BDG test levels in 120 immunocompetent and uninfected pediatric patients;[47] the median age was 9.2 years (range, 7 months to 8 years). The median BDG level was 32 pg/mL and the mean value was 68 pg/mL. The mean values did not vary significantly by age or gender. These observations suggested that normal mean BDG values in children are in general higher than those from adults tested previously (68 pg/mL versus 48 pg/mL for children and adults, respectively), raising questions about the critical cut-off values of BDG in children. These two parameters—lack of data and possible higher normal BDG values in children—have led to the absence of recommendations of BDG testing in children.

Imaging studies

To date, the standard imaging technique for the detection of IA remains the CT scan. Characteristic CT findings for IA in adults include pulmonary nodules, halo sign, air crescent sign, and cavitation.[48,49] In adult series studies, approximately 50% of cases show cavitation, with air crescent formation in 40%. These findings constitute basic clinical criteria in the revised EORTC/MSG definitions of invasive fungal disease.[38] However, the main drawbacks of CT scan are the limited specificity and predictive value. Previous papers reported that CT scans do not always allow the differentiation of pathogenic fungi. PET scan with CT imaging has been proposed as an alternative diagnostic procedure, which increases the specificity of detecting IA, since it provides a detailed and more functional information of findings; MRI findings are helpful in detecting cases of *Aspergillus* osteomyelitis and CNS involvement.[50] High T2 signal found in the cortex or subcortical white matter and, sometimes, the presence of hemorrhage representing infarcts are indicative of cerebral aspergillosis. Furthermore, up to 30% of adult patients with pulmonary aspergillosis may present multiple organ involvement, and for that reason multisystem radiological evaluation of high risk patients is required.

The main problem in the pediatric setting regarding imaging techniques is primarily the lack of specific lesions for the detection of IA.[51–55] Radiographic findings in children are often nonspecific, especially in the younger age group (<5 years). Second, the limited data on imaging studies are focused only on immunocompromised children with underlying malignancies; information in other pediatric patient groups is lacking.[55] In a 10-year study evaluating radiographic findings of IA in children, a large variation of the lesions, which were often nonspecific, was reported.[52] Furthermore, the cavi-

tation rate was found to be less common in children than in adult patients. Finally, the utility of a CT scan was limited to detecting multiple lesions and involvement of the chest wall.

According to the 2011 ECIL 4 Group recommendations, in high-risk children with febrile neutropenia >96 h or with focal clinical findings, imaging studies (lung CT scan or adequate imaging of the symptomatic region) should be performed.[38] Even atypical pulmonary infiltrates (e.g., fluffy masses) may support the diagnosis of invasive pulmonary aspergillosis in pediatric high risk patients. At least, a further diagnostic work-up (e.g., BAL, biopsy) should be considered in this patient group and mold-active antifungal treatment should be promptly initiated.

Novel imaging studies

In the recent years, novel sensitive radiotracers have been identified for the imaging of IA, such as antimicrobial peptides, antifungal agents, and chitin-specific agents.[56] Two siderophores, triacetylfusarinine (TAFC) and ferrioxamine E (FOXE), have been labeled with gallium-68 and used for PET imaging of *A. fumigatus* infection in rats. Based on results, both tracers revealed high metabolic stability and a favorable biodistribution.[57] A peptide (c(CGGRLGPFC)-NH_2) labeled with indium-111 was also specifically bound on *Aspergillus* hyphae.[58] These novel sensitive radiotracers constitute future candidate technologies for IA imaging, but further investigation is required to assess the *in vivo* biodistribution and safety in human clinical trials.

Conclusions

Prompt diagnosis of IA in the pediatric setting remains a major challenge. Awareness of the optimal diagnostic procedures solely or in combination is mandatory for prompt detection of the etiologic fungal species. Standard and newer diagnostic methods of IA are extensively evaluated in adults, but these data cannot be simply extrapolated to pediatric patients.

In summary, the GM test can be used for children with caution, the utility of the β-1,3-D-glucan needs further research for children, and imaging, particularly the CT scan, is a useful but nonspecific tool, whereas molecular markers such as PCR present the same problems and difficulties as in adults. While progress has been achieved in terms of GM and certain recommendations have been made, further research is needed for the validation of newer diagnostic procedures in pediatric patients.

Conflicts of interest

The authors declare no conflicts of interest.

References

1. Roilides, E. 2006. Early diagnosis of invasive aspergillosis in infants and children. *Med. Mycol.* **44:** 199–205.
2. Steinbach, W.J. 2005. Pediatric aspergillosis: disease and treatment differences in children. *Pediatr. Infect. Dis. J.* **24:** 358–364.
3. Ozen, M. & N.O. Dundar. 2011. Invasive aspergillosis in children with hematological malignancies. *Expert Rev. Anti. Infect. Ther.* **9:** 299–306.
4. Salman, N. *et al.* 2011. Invasive aspergillosis in hematopoietic stem cell and solid organ transplantation. *Expert Rev. Anti. Infect. Ther.* **9:** 307–315.
5. Shetty, D. *et al.* 1997. Invasive aspergillosis in human immunodeficiency virus-infected children. *Pediatr. Infect. Dis. J.* **16:** 216–221.
6. Alsultan, A. *et al.* 2006. Chronic granulomatous disease presenting with disseminated intracranial aspergillosis. *Pediatr. Blood Cancer.* **47:** 107–110.
7. Guillot, M. *et al.* 2003. Aspergillosis in cystic fibrosis patients. *Arch. Pediatr.* **10**(Suppl 5): 588s–591s.
8. Schwartz, D.A., M. Jacquette & H.S. Chawla. 1988. Disseminated neonatal aspergillosis: report of a fatal case and analysis of risk factors. *Pediatr. Infect. Dis. J.* **7:** 349–353.
9. Dinleyici, E.C. 2011. Pediatric invasive fungal infections: realities, challenges, concerns, myths and hopes. *Expert Rev. Anti. Infect. Ther.* **9:** 273–274.
10. Hope, W.W., T.J. Walsh & D.W. Denning. 2005. Laboratory diagnosis of invasive aspergillosis. *Lancet Infect. Dis.* **5:** 609–622.
11. Mengoli, C. *et al.* 2009. Use of PCR for diagnosis of invasive aspergillosis: systematic review and meta-analysis. *Lancet Infect. Dis.* **9:** 89–96.
12. Zou, M. *et al.* 2011. Systematic review and meta-analysis of detecting galactomannan in bronchoalveolar lavage fluid for diagnosing invasive aspergillosis. *PLoS One* **7:** e43347.
13. Johnson, G.L. *et al.* 2012. A MIQE-compliant real-time PCR assay for Aspergillus detection. *PLoS One* **7:** e40022.
14. Klingspor, L. & J. Loeffler. 2009. Aspergillus PCR formidable challenges and progress. *Med. Mycol.* **47**(Suppl 1): S241–S247.
15. McCulloch, E. *et al.* 2012. Antifungal treatment affects the laboratory diagnosis of invasive aspergillosis. *J. Clin. Pathol.* **65:** 83–86.
16. Halliday, C. *et al.* 2006. Role of prospective screening of blood for invasive aspergillosis by polymerase chain reaction in febrile neutropenic recipients of haematopoietic stem cell transplants and patients with acute leukaemia. *Br. J. Haematol.* **132:** 478–486.
17. Hummel, M. *et al.* 2009. Detection of Aspergillus DNA by a nested PCR assay is able to improve the diagnosis of invasive aspergillosis in paediatric patients. *J. Med. Microbiol.* **58:** 1291–1297.
18. Shi, L.N. *et al.* 2012. Immunoproteomics based identification of thioredoxin reductase GliT and novel Aspergillus fumigatus antigens for serologic diagnosis of invasive aspergillosis. *BMC Microbiol.* **12:** 11.
19. Kumar, A. *et al.* 2011. Identification of virulence factors and diagnostic markers using immunosecretome of Aspergillus fumigatus. *J. Proteomics* **74:** 1104–1112.
20. Gautam, P. *et al.* 2007. Identification of novel allergens of Aspergillus fumigatus using immunoproteomics approach. *Clin. Exp. Allergy.* **37:** 1239–1249.
21. Patterson, T.F. 2011. Clinical utility and development of biomarkers in invasive aspergillosis. *Trans. Am. Clin. Climatol. Assoc.* **122:** 174–183.
22. Bergeron, A. *et al.* 2011. Prospective evaluation of clinical and biological markers to predict the outcome of invasive pulmonary aspergillosis in hematological patients. *J. Clin. Microbiol.* **50:** 823–830.
23. Verweij, P.E. 2005. Advances in diagnostic testing. *Med. Mycol.* **43**(Suppl 1): S121–S124.
24. Rohrlich, P. *et al.* 1996. Prospective sandwich enzyme-linked immunosorbent assay for serum galactomannan: early predictive value and clinical use in invasive aspergillosis. *Pediatr. Infect. Dis. J.* **15:** 232–237.
25. Sulahian, A. *et al.* 2001. Value of antigen detection using an enzyme immunoassay in the diagnosis and prediction of invasive aspergillosis in two adult and pediatric hematology units during a 4-year prospective study. *Cancer* **91:** 311–318.
26. Herbrecht, R. *et al.* 2002. Aspergillus galactomannan detection in the diagnosis of invasive aspergillosis in cancer patients. *J. Clin. Oncol.* **20:** 1898–1906.
27. Challier, S. *et al.* 2004. Development of a serum-based Taqman real-time PCR assay for diagnosis of invasive aspergillosis. *J. Clin. Microbiol.* **42:** 844–846.
28. El-Mahallawy, H.A. *et al.* 2006. Evaluation of pan-fungal PCR assay and Aspergillus antigen detection in the diagnosis of invasive fungal infections in high risk paediatric cancer patients. *Med. Mycol.* **44:** 733–739.
29. Hovi, L. *et al.* 2007. Prevention and monitoring of invasive fungal infections in pediatric patients with cancer and hematologic disorders. *Pediatr. Blood Cancer.* **48:** 28–34.
30. Steinbach, W.J. *et al.* 2007. Prospective Aspergillus galactomannan antigen testing in pediatric hematopoietic stem cell transplant recipients. *Pediatr. Infect. Dis. J.* **26:** 558–564.
31. Hayden, R. *et al.* 2008. Galactomannan antigenemia in pediatric oncology patients with invasive aspergillosis. *Pediatr. Infect. Dis. J.* **27:** 815–819.
32. Armenian, S.H. *et al.* 2009. Prospective monitoring for invasive aspergillosis using galactomannan and polymerase chain reaction in high risk pediatric patients. *J. Pediatr. Hematol. Oncol.* **31:** 920–926.
33. Castagnola, E. *et al.* 2010. Performance of the galactomannan antigen detection test in the diagnosis of invasive aspergillosis in children with cancer or undergoing haemopoietic stem cell transplantation. *Clin. Microbiol. Infect.* **16:** 1197–1203.
34. de Mol, M. *et al.* 2012. Diagnosis of invasive pulmonary aspergillosis in children with bronchoalveolar lavage galactomannan. *Pediatr. Pulmonol.* PMID: 22949309.

35. Verweij, P.E. *et al.* 2000. Failure to detect circulating Aspergillus markers in a patient with chronic granulomatous disease and invasive aspergillosis. *J. Clin. Microbiol.* **38:** 3900–3901.

36. Mortensen, K.L. *et al.* 2011. Successful management of invasive aspergillosis presenting as pericarditis in an adult patient with chronic granulomatous disease. *Mycoses* **54:** e233–236.

37. Pfeiffer, C.D., J.P. Fine & N. Safdar. 2006. Diagnosis of invasive aspergillosis using a galactomannan assay: a meta-analysis. *Clin. Infect. Dis.* **42:** 1417–1427.

38. 4th European Conference on Infections in Leukemia ECIL 4. 2012. Pediatric Group Considerations for Fungal Diseases and Antifungal Treatment in Children Pediatric Guidelines for antifungals. http://www.leukemia-net.org/content/treat_research/supportive_care/standards_sop_and_recommendations/index_eng.html

39. Lehrnbecher, T. & A.H. Groll. 2011. Invasive fungal infections in the pediatric population. *Expert Rev. Anti-infect. Ther.* **9:** 275–278.

40. Desai, R., L.A. Ross & J.A. Hoffman. 2009. The role of bronchoalveolar lavage galactomannan in the diagnosis of pediatric invasive aspergillosis. *Pediatr. Infect. Dis. J.* **28:** 283–286.

41. Viscoli, C. *et al.* 2002. Aspergillus galactomannan antigen in the cerebrospinal fluid of bone marrow transplant recipients with probable cerebral aspergillosis. *J. Clin. Microbiol.* **40:** 1496–1499.

42. Roilides, E. *et al.* 2003. Cerebral aspergillosis in an infant with corticosteroid-resistant nephrotic syndrome. *Pediatr. Nephrol.* **18:** 450–453.

43. Karageorgopoulos, D.E. *et al.* 2011. Accuracy of beta-d-glucan for the diagnosis of Pneumocystis jirovecii pneumonia: a meta-analysis. *Clin. Microbiol. Infect.* PMID: 22329494.

44. Oz, Y. & N. Kiraz. 2011. Diagnostic methods for fungal infections in pediatric patients: microbiological, serological and molecular methods. *Expert Rev. Anti-infect. Ther.* **9:** 289–298.

45. de Pauw, B.E. & J.J. Picazo. 2008. Present situation in the treatment of invasive fungal infection. *Int. J. Antimicrob. Agents.* **32**(Suppl 2)**:** S167–S171.

46. Mularoni, A. *et al.* 2011. High Levels of beta-D-glucan in immunocompromised children with proven invasive fungal disease. *Clin. Vaccine Immunol.* **17:** 882–883.

47. Smith, P.B. *et al.* 2007. Quantification of 1,3-beta-D-glucan levels in children: preliminary data for diagnostic use of the beta-glucan assay in a pediatric setting. *Clin. Vaccine Immunol.* **14:** 924–925.

48. Caillot, D. *et al.* 1997. Improved management of invasive pulmonary aspergillosis in neutropenic patients using early thoracic computed tomographic scan and surgery. *J. Clin. Oncol.* **15:** 139–147.

49. Heussel, C.P. *et al.* 1999. Multiple renal aspergillus abscesses in an AIDS patient: contrast-enhanced helical CT and MRI findings. *Eur. Radiol.* **9:** 616–619.

50. Ashdown, B.C., R.D. Tien & G.J. Felsberg. 1994. Aspergillosis of the brain and paranasal sinuses in immunocompromised patients: CT and MR imaging findings. *AJR Am. J. Roentgenol.* **162:** 155–159.

51. Tragiannidis, A. *et al.* 2012. Invasive aspergillosis in children with acquired immunodeficiencies. *Clin. Infect. Dis.* **54:** 258–267.

52. Thomas, K.E. *et al.* 2003. The radiological spectrum of invasive aspergillosis in children: a 10-year review. *Pediatr. Radiol.* **33:** 453–460.

53. Burgos, A. *et al.* 2008. Pediatric invasive aspergillosis: a multicenter retrospective analysis of 139 contemporary cases. *Pediatrics* **121:** e1286–e1294.

54. Taccone, A. *et al.* 1993. CT of invasive pulmonary aspergillosis in children with cancer. *Pediatr. Radiol.* **23:** 177–180.

55. Archibald, S. *et al.* 2001. Computed tomography in the evaluation of febrile neutropenic pediatric oncology patients. *Pediatr. Infect. Dis. J.* **20:** 5–10.

56. Lupetti, A. *et al.* 2011. Radiotracers for fungal infection imaging. *Med. Mycol.* **49**(Suppl 1)**:** S62–S69.

57. Petrik, M. *et al.* 2012. Preclinical evaluation of two 68Ga-siderophores as potential radiopharmaceuticals for Aspergillus fumigatus infection imaging. *Eur. J. Nucl. Med. Mol. Imaging* **39:** 1175–1183.

58. Yang, Z. *et al.* 2009. Gamma scintigraphy imaging of murine invasive pulmonary aspergillosis with a (111)In-labeled cyclic peptide. *Nucl. Med. Biol.* **36:** 259–266.

Ann. N.Y. Acad. Sci. ISSN 0077-8923

ANNALS OF THE NEW YORK ACADEMY OF SCIENCES
Issue: *Advances Against Aspergillosis*

Azole resistance in *Aspergillus*: global status in Europe and Asia

Sevtap Arikan-Akdagli

Department of Medical Microbiology, Hacettepe University Medical School, Ankara, Turkey

Address for correspondence: Prof. Sevtap Arikan-Akdagli, M.D., Hacettepe University Medical School, Department of Medical Microbiology, 06100 Ankara, Turkey. sevtap.arikan@gmail.com

Azole-resistant strains of *Aspergillus* have been reported from European and Asian countries at varying frequencies. Based on the limited rates of isolation of *Aspergillus* from clinical samples in routine practice and the limited number of the screening studies carried out so far, the true prevalence of triazole resistance and the rate of multiazole-resistant strains remain partly unknown. Also, available data are mostly for *A. fumigatus* (complex), thus the situation for non-*fumigatus* Aspergilli is less clear. In general, exposure of *Aspergillus* to antifungal agents via medical or environmental (agricultural) use of these compounds appears to have the possible major impact on acquisition of triazole resistance. Azole resistance in *Aspergillus* remains to be further elucidated by continued surveillance studies. Based on the possible association with agricultural azole use, environmental sampling appears significant as well.

Keywords: *Aspergillus*; azole; resistance; antifungal susceptibility testing

Introduction

Azole resistance in the genus *Aspergillus* initially drew attention following a report of itraconazole-resistant *Aspergillus fumigatus* strains in 1997. This report focused on three clinical isolates (AF72, AF90, AF91) with high itraconazole minimum inhibitory concentrations (MICs) (>16 µg/mL), validation of the *in vitro* resistance by a neutropenic murine model, and investigation of the possible resistance mechanisms (changes in sterol 14α-demethylase and alteration in a membrane transporter). Of note, these strains were from California and isolated in 1980s.[1] Following the identification of these resistant isolates, optimal test parameters have been proposed for both broth and agar dilution methods for generation of reproducible MICs and prediction of clinical outcome.[2]

Proposed breakpoints and epidemiological cut-off values for determination of azole resistance in *Aspergillus*

To facilitate the analysis and interpretation of the MIC data for azoles, CLSI (Clinical and Laboratory Standards Institute) and EUCAST (The European Committee on Antimicrobial Susceptibility Testing)

MIC breakpoints for itraconazole, voriconazole, and posaconazole against *A. fumigatus* were initially proposed in 2009. Using these breakpoints, resistance was defined as MICs of >2, >2, >0.5 µg/mL, for itraconazole, voriconazole, and posaconazole, respectively.[3]

The issue of determination and validation of epidemiological cut-off values (ECOFFs/ECVs) and antifungal clinical breakpoints for *Aspergillus* is still dynamic and now includes further analysis by using both CLSI and EUCAST methodologies. ECOFFs/ECVs are the upper MIC values defining the wild-type distributions, and antifungal clinical breakpoints are the MIC values indicating likely clinical response. Using CLSI M38-A2 methodology,[4] ECVs have been proposed for itraconazole, posaconazole, and voriconazole against strains of six *Aspergillus* spp. (*A. fumigatus*, *A. flavus*, *A. terreus*, *A. niger*, *A. nidulans*, and *A. versicolor*[5]) (Table 1). Likewise, ECOFFs have been proposed for the EUCAST method and for testing *A. fumigatus*, *A. flavus*, *A. terreus*, *A. niger*, and *A. nidulans* (Table 1). Based on the available microbiological data and clinical experience, EUCAST species-specific clinical breakpoints have also been proposed for some azole-species combinations[6–9]

doi: 10.1111/j.1749-6632.2012.06815.x
Ann. N.Y. Acad. Sci. 1272 (2012) 9–14 © 2012 New York Academy of Sciences.

Table 1. ECVs (μg/mL)/ECOFFs (mg/L) for azoles against *Aspergillus* species

Species	Methodology	
Azole	CLSI (ECVs)	EUCAST (ECOFFs)
A. fumigatus		
Itraconazole	1	1
Voriconazole	1	1
Posaconazole	0.5 (0.25)[*]	0.25
A. flavus		
Itraconazole	1	1
Voriconazole	1	2
Posaconazole	0.25	0.5
A. terreus		
Itraconazole	1	0.5[**]
Voriconazole	1	2
Posaconazole	0.5	0.25
A. niger		
Itraconazole	2	4
Voriconazole	2	2
Posaconazole	0.5	0.5
A. nidulans		
Itraconazole	1	1
Voriconazole	2	1
Posaconazole	1	0.5
A. versicolor		
Itraconazole	2	ND
Voriconazole	2	ND
Posaconazole	1[***]	ND

Data from Refs. 5–8.
[*]Eyeball (statistically calculated) ECVs.
[**]Has recently been lowered from 1 to 0.5 mg/L since 0.5 appeared to better describe the wild-type distribution.[10]
[***]Statistically calculated ECVs. ND, not determined.

(Table 2). These ECOFFs and breakpoint values are under current discussion and evaluation.

Surveillance studies for detection of azole-resistant *Aspergillus* strains

Nationwide, single-center national, and international surveillance studies have been carried out to explore the possible existence and extent of azole resistance in *Aspergillus* strains. There have been reports of resistant strains in various countries, including France, Spain, Sweden, and the United Kingdom, as well as the United States and Canada.[11] Be-

ing the most commonly isolated species in routine practice, *A. fumigatus* has frequently remained as the most commonly tested species as well in *in vitro* susceptibility studies.[12,13]

Nationwide and single-center national studies

The Netherlands
A nationwide multicenter survey carried out in the Netherlands including 170 strains of *A. fumigatus* isolated over 53 years (1945–1998) showed the existence of three strains with high itraconazole MICs (64 μg/mL) and no strains with high voriconazole MICs.[14] The issue gained further importance by the demonstration of the gradual increase in itraconazole resistance (1.7 to 6% within the time period starting after 1999 till 2007) in *A. fumigatus* strains in Nijmegen, the Netherlands.[15] In the noted study by Snelders *et al.*, the isolates growing in presence of 8 μg/mL of itraconazole were tested for their *in vitro* susceptibilities against voriconazole, ravuconazole, and posaconazole as well by using CLSI M38-A microdilution methodology.[4] Importantly and for itraconazole-resistant strains, elevated MICs and reduced susceptibilities were detected for voriconazole, posaconazole, and ravuconazole as well.[15]

Another noteworthy observation in the study by Snelders *et al.*[15] was the spread of a single azole resistance mechanism. As investigated in multicenter strains in the Netherlands, as well as in strains from six other European countries (Belgium, France, Greece, Norway, Sweden, and the United Kingdom), the dominant resistance mechanism was detected to be a combination of a substitution of leucine 98 for histidine in the *cyp51A* gene and two copies of a 34-bp sequence in tandem in the gene promoter (TR/L98H). This observation has verified the existence of the resistance mechanism of TR/L98H in the Netherlands and Norway, in addition to other reports from Belgium, France, Spain, Sweden, and the United Kingdom.[15]

In addition, there were findings suggesting an environmental source rather than a person-to-person transmission for these resistant isolates. The related evidence was the dominance of the resistance mechanism in clinical isolates from epidemiologically unrelated patients, its existence in some isolates from azole-naive patients, and shorter genetic distances in microsatellite typing between the TR/L98H isolate compared to that for other azole-resistant

Table 2. Clinical breakpoints (mg/L) proposed by EUCAST for triazoles against *Aspergillus* spp.

| Antifungal drug | Species | | | | | | | | | | Non-species related breakpoints | |
| | *A. fumigatus* | | *A. flavus* | | *A. nidulans* | | *A. niger* | | *A. terreus* | | | |
	S ≤	R >	S ≤	R >	S ≤	R >	S ≤	R >	S ≤	R >	S ≤	R >
Itraconazole	1	2	1	2	1	2	IE	IE	1	2	IE	IE
Posaconazole	0.12	0.25	IE	IE	IE	IE	IE	IE	0.12	0.25	IE	IE
Voriconazole	IP	IP	IP	IP	IP	IP	IP	IP	IP	IP	IP	IP

Data adapted from Ref. 9.
S, susceptible; R, resistant; IE, insufficient evidence; IP, in preparation.

and azole-susceptible strains. These findings suggested the existence of a spreading azole mechanism among strains of *A. fumigatus* and the need for screening for azole resistance. Also, the question of whether the agricultural use of azole-containing fungicides might have been the source for the development of azole resistance in the environment was raised.[15]

The prospective nationwide multicenter study from the Netherlands (June 2007–January 2009) showed an itraconazole resistance prevalence of 5.3% (range: 0.8–9.5%) for *A. fumigatus*, suggesting widespread multiazole resistance in the country. Importantly, 64% of the patients who harbored an azole-resistant strain were found to be azole-naive.[16]

The United Kingdom
Azole resistance was explored in clinical *A. fumigatus* strains that were received in the Regional Mycology Laboratory in Manchester, UK from 1997 to 2007. The frequency of itraconazole resistance in these strains was reported as 5%; 65% and 74% of them being cross-resistant to voriconazole and posaconazole, respectively (ECOFFs used for analysis: >2, >2, >0.5 μg/mL for itraconazole, voriconazole, and posaconazole, respectively). Importantly, the first itraconazole-resistant strain was detected to be among the strains isolated in 1999.[17]

Japan
Using the epidemiological cut-off values of 1, 0.5, and 1 μg/mL for itraconazole, posaconazole, and voriconazole, respectively, and for the clinical *A. fumigatus* strains isolated in Nagasaki University Hospital, Nagasaki, Japan, the percentages of non-

wild-type isolates were found as 7.1, 2.6, and 4.1 for the denoted azoles.[18]

India
Two of 103 clinical *A. fumigatus* strains isolated at University of Delhi, India in years 2005 to 2010 were shown to have high MICs for azoles (rate of resistance: 1.9%). The MICs of these strains for itraconazole, voriconazole, posaconazole, and isavuconazole were > 16, 2, 2, and 8 μg/mL, respectively. The strains were isolated from azole-naive patients with chronic respiratory disease.[19]

International surveillance studies

Two international surveillance studies have so far been carried out for detection of azole resistance in *Aspergillus*. The prospective international surveillance study SCARE included strains from 23 participating centers and 20 countries. The prevalence of azole resistance in *Aspergillus* section *fumigati* was found to be 0–4.2% per center and resistant strains were detected in isolates from Austria, Australia, Belgium, Denmark, France, Italy, the Netherlands, Spain, Sweden, Switzerland, and the United Kingdom.[20]

In the ARTEMIS global surveillance study, isolates from 62 medical centers worldwide were tested. Among the 2008–2009 *A. fumigatus* strains, 5.8% generated elevated MICs to one or more triazoles. Most (82.3%) of these resistant strains were from Hangzhou, China and the rest were from centers in the Czech Republic, Portugal, Brazil, and the United States.[21,22]

Azole resistance in strains of *A. fumigatus* isolated from patients with specific underlying disorders

The azole resistance rates in *Aspergillus* have been explored in specific patient settings, including hematological malignancies, cystic fibrosis, and chronic pulmonary disease due to *Aspergillus*.

A prospective 4-year study, including *A. fumigatus* strains isolated from a French cohort of patients treated for hematological malignancies, revealed itraconazole resistance in only 1 of 118 strains (0.85%).[23] Azole resistance in *A. fumigatus* strains isolated from patients with csytic fibrosis was reported from France and Denmark. Itraconazole MICs of ≥ 2 µg/mL were detected in 4.6% of patients with cystic fibrosis at Cochin University Hospital, France. Importantly, azole-resistant *A. fumigatus* strains were detected in 20% of subjects who received itraconazole within the previous three years.[24] The report from Denmark, on the other hand, revealed the existence of azole-non-susceptible or azole-resistant *A. fumigatus* strains in 6 of 133 cystic fibrosis patients (4.5%). All six patients had been exposed to an azole previously—46–278 weeks prior to detection of the resistant isolate and longer than that for patients with no mold isolation or those who harbored azole-susceptible *Aspergillus* isolates.[25] These findings further emphasized the well-known association between azole resistance and previous exposure.

Based on the common occurrence of very low organism burden and a negative culture, molecular studies have been carried out for detection of azole resistance in patients with chronic pulmonary aspergillosis. Triazole resistance in *A. fumigatus* strains from lungs of patients with chronic fungal disease was studied in samples that were culture negative but polymerase chain reaction (PCR) positive. Importantly, triazole-resistance mutations (L98H/TR and M220) were detected in as high as 55% of these samples.[26]

Is azole resistance increasing in strains of *A. fumigatus*?

Increase in rate of azole resistance in *A. fumigatus* strains has so far been reported only from some centers. The single-center surveillance study in the Netherlands (screening of 1912 strains isolated from 1994 to 2007) showed an increase in itraconazole resistance after 1999.[15] Similarly, among the isolates submitted to Regional Mycology Laboratory, Manchester, UK, an increase in azole resistance has been detected since 2004,[17] and the continuation of this surveillance study in 2008 and 2009 showed a continued increase in azole resistance after 2007 as well.[27]

In contrast to the reports from the Netherlands and United Kingdom, no increase has been observed in voriconazole MICs in the post- versus pre-voriconazole era in Spain.[28] Similarly, a part of the ARTEMIS study that included isolates of nine years (from 2001 to 2009) showed no consistent trend toward decreased susceptibility to any triazole by *A. fumigatus*.[29] These results indicate that an increase in azole resistance in *A. fumigatus* may constitute a current problem at least in some centers, and surveillance studies need to be carried out in each center to reveal a possible trend toward an increase in resistance.

Data available for non-*fumigatus* *Aspergillus*

As noted previously, most available data on azole resistance cover only the *A. fumigatus* complex. Except for these studies, there is a published report on azole resistance in the *A. niger* complex. Interestingly and overall, 52% of the isolates included in this study were itraconazole resistant. Importantly, the rates of itraconazole resistance varied remarkably from one clade to other (100, 90, 36, and 33% in *A. acidus*, *A. tubingensis*, *A. awamori*, and *A. niger*, respectively). Data obtained in this study indicated that itraconazole resistance was common in the *A. niger* complex but cyp51A mutations might not be playing an important role. Azole cross-resistance, on the other hand, was unusual among the strains of the *A. niger* complex.[30] Further data are awaited on rates of azole resistance among strains belonging to *A. fumigatus* and *A. niger*, as well as other *Aspergillus* species complexes.

Environmental studies

The possible role of agricultural use of azole-containing fungicides in emergence of azole resistance in *Aspergillus* strains evoked studies on environmental sources of resistance. In one of these environmental studies in the Netherlands, itraconazole-resistant *A. fumigatus* strains were cultivated from indoor hospital environment, soil obtained from flower beds, commercial compost, leaves, and seeds. Remarkably, cross-resistance was observed to voriconazole and posaconazole, as to well as to azole fungicides (metcunazole, tebuconazole). Also, resistant environmental and clinical isolates were found to be genetically clustered and apart from nonresistant strains. These results suggest that colonization with azole-resistant isolates from the environment is possible.[31]

The existence of azole-resistant *Aspergillus* in environmental samples was explored in Austria, Denmark, and Spain. Multi-azole-resistant *A. fumigatus* strains were isolated from soil samples in Denmark, and an *A. lentulus* strain with high voriconazole MIC was isolated from the soil sample in Spain.[32]

Conclusions

Azole resistance in *A. fumigatus* appears to have emerged at varying rates in European and Asian countries. Further studies on clinical and environmental isolates belonging to various *Aspergillus* species complices may provide more insight and elucidation for the true prevalence, mechanism, and clinical impact of azole resistance in *Aspergillus*.

Conflicts of interest

The author does not have any potential conflicts of interests related particularly to this paper.

Otherwise, she has received investigator initiated research grant support from Pfizer and speaker honoraria from Merck and Pfizer.

References

1. Denning, D.W., K. Venkateswarlu, K.L. Oakley, *et al.* 1997. Itraconazole resistance in *Aspergillus fumigatus*. *Antimicrob. Agents Chemother*. **41:** 1364–1368.

2. Denning, D.W., S.A. Radford, K.L. Oakley, *et al.* 1997. Correlation between in-vitro susceptibility testing to itraconazole and in-vivo outcome of *Aspergillus fumigatus* infection. *J. Antimicrob. Chemother*. **40:** 401–414.

3. Verweij, P.E., S.J. Howard, W.J.G. Melchers & D.W. Denning. 2009. Azole-resistance in *Aspergillus*: proposed nomenclature and breakpoints. *Drug Resist. Updates*. **12:** 141–147.

4. CLSI. 2008. *Reference Method for Broth Dilution Antifungal Susceptibility Testing of Filamentous Fungi; Approved Standard*. 2nd ed. CLSI Document M38-A2. Clinical and Laboratory Standards Institute. Wayne, PA.

5. Espinel-Ingroff, A., D.J. Diekema, A. Fothergill, *et al.* 2010. Wild-type MIC distributions and epidemiological cut-off values for the triazoles and six *Aspergillus* spp. for the CLSI broth microdilution method (M38-A2 document). *J. Clin. Microbiol*. **48:** 3251–3257.

6. Itraconazole and *Aspergillus* spp. Rationale for the EUCAST clinical breakpoints, version 1.0. 11 January 2012.

7. Posaconazole and *Aspergillus* spp. Rationale for the EUCAST clinical breakpoints, version 1.0. 17 January 2012.

8. Voriconazole. Rationale for the EUCAST clinical breakpoints, version 1.0. March 2012.

9. EUCAST Antifungal Breakpoints: www.eucast.org/fileadmin/src/media/PDFs/EUCAST_files/AFST/Antifungal_breakpoints_v_4.1.pdf. Accessed October 29, 2012.

10. EUCAST Comments on proposed itraconazole breakpoints for Aspergillus spp. Available at: www.eucast.org/fileadmin/src/media/PDFs/EUCAST_files/AFST/EUCAST_Itra_consultation_comments_and_responses_f.pdf. Accessed January 11, 2012

11. Denning, D.W. & D.S. Perlin. 2011. Azole resistance in *Aspergillus*: a growing public health menace. *Future. Microbiol*. **6:** 1229–1232.

12. Pfaller, M., L. Boyken, R. Hollis, J. Kroeger, *et al.* 2011. Comparison of the broth microdilution methods of the European Committee on Antimicrobial Susceptibility Testing and the Clinical and Laboratory Standards Institute for testing itraconazole, posaconazole, and voriconazole against *Aspergillus* isolates. *J. Clin. Microbiol*. **49:** 1110–1112.

13. Arikan, S., B. Sancak, S. Alp, *et al.* 2008. Comparative in vitro activities of posaconazole, voriconazole, itraconazole, and amphotericin B against *Aspergillus* and *Rhizopus*, and synergy testing for *Rhizopus*. *Med. Mycol*. **46:** 567–573.

14. Verweij, P.E., D.T. Te Dorsthorst, A.J. Rijs, *et al.* 2002. Nationwide survey of in vitro activities of itraconazole and voriconazole against clinical *Aspergillus fumigatus* isolates cultured between 1945 and 1998. *J. Clin. Microbiol*. **40:** 2648–2650.

15. Snelders, E., H.A. van der Lee, J. Kuijpers, *et al.* 2008. Emergence of azole resistance in *Aspergillus fumigatus* and spread of a single resistance mechanism. *PLoS Med*. **5:** e219.

16. Van der Linden, J.W., E. Snelders, G.A. Kampinga, *et al.* 2011. Clinical implications of azole resistance in *Aspergillus fumigatus*, The Netherlands, 2007–2009. *Emerging Infect. Dis*. **17:** 1846–1854.

17. Howard, S., D. Cerar, M.J. Anderson, *et al.* 2009. Frequency and evolution of azole resistance in *Aspergillus fumigatus* associated with treatment failure. *Emerg. Infect. Dis*. **15:** 1068–1076.

18. Tashiro, M., K. Izumikawa, A. Minematsu, *et al.* 2012. Antifungal susceptibilities of *Aspergillus fumigatus* clinical isolates obtained in Nagasaki, Japan. *Antimicrob. Agents Chemother*. **56:** 584–587.

19. Chowdhary, A., S. Kathuria, H.S. Randhawa, *et al.* 2012. Isolation of multiple-triazole-resistant *Aspergillus fumigatus* strains carrying the TR/L98H mutations in the cyp51A gene in India. *J. Antimicrob. Chemother*. **67:** 362–366.

20. Van der Linden, J., M.C. Arendrup, P.E. Verweij, SCARE Network. 2011. Prospective international surveillance of azole resistance in *Aspergillus fumigatus*. 51st Interscience Conference on Antimicrobial Agents and Chemotherapy, September 17–20, Chicago USA, M-490.

21. Pfaller, M.A., S.A. Messer, L. Boyken, *et al.* 2008. In vitro survey of triazole cross-resistance among more than 700 clinical isolates of *Aspergillus* species. *J. Clin. Microbiol*. **46:** 2568–2572.

22. Lockhart, S.R., J.P. Frade, K.A. Etienne, *et al.* 2011. Azole resistance in *Aspergillus fumigatus* isolates from the ARTEMIS Global Surveillance Study is primarily due to the TR/L98H mutation in the *cyp51A* gene. *Antimicrob. Agents Chemother.* **55:** 4465–4468.

23. Alanio, A., E. Sitterlé, M. Liance, *et al.* 2011. Low prevalence of resistance to azoles in *Aspergillus fumigatus* in a French cohort of patients treated for haematological malignancies. *J. Antimicrob. Chemother.* **66:** 371–374.

24. Burgel, P.R., M.T. Baixench, M. Amsellem, *et al.* 2012. High prevalence of azole-resistant *Aspergillus fumigatus* in adults with cystic fibrosis exposed to itraconazole. *Antimicrob. Agents Chemother.* **56:** 869–874.

25. Mortensen, K.L., R.H. Jensen, H.K. Johansen, *et al.* 2011. *Aspergillus* species and other molds in respiratory samples from patients with cystic fibrosis: a laboratory-based study with focus on *Aspergillus fumigatus* azole resistance. *J. Clin. Microbiol.* **49:** 2243–2251.

26. Denning, D.W., S. Park, C. Lass-Florl, *et al.* 2011. High-frequency triazole resistance found in nonculturable *Aspergillus fumigatus* from lungs of patients with chronic fungal disease. *Clin. Infect. Dis.* **52:** 1123–1129.

27. Bueid, A., S.J. Howard, C.B. Moore, *et al.* 2010. Azole antifungal resistance in *Aspergillus fumigatus*: 2008 and 2009. *J. Antimicrob. Chemother.* **65:** 2116–2118.

28. Guinea, J., S. Recio, T. Peláez, *et al.* 2008. Clinical isolates of *Aspergillus* species remain fully susceptible to voriconazole in the post-voriconazole era. *Antimicrob. Agents Chemother.* **52:** 3444–3446.

29. Pfaller, M., L. Boyken, R. Hollis, *et al.* 2011. Use of epidemiological cut-off values to examine 9-year trends in susceptibility of *Aspergillus* species to the triazoles. *J. Clin. Microbiol.* **49:** 586–590.

30. Howard, S.J., E. Harrison, P. Bowyer, *et al.* 2011. Cryptic species and azole resistance in the *Aspergillus niger* complex. *Antimicrob. Agents Chemother.* **55:** 4802–4809.

31. Snelders, E., R.A. Huis In 't Veld, A.J. Rijs, *et al.* 2009. Possible environmental origin of resistance of *Aspergillus fumigatus* to medical triazoles. *Appl. Environ. Microbiol.* **75:** 4053–4057.

32. Mortensen, K.L., E. Mellado, C. Lass-Flörl, *et al.* 2010. Environmental study of azole-resistant *Aspergillus fumigatus* and other Aspergilli in Austria, Denmark, and Spain. *Antimicrob. Agents Chemother.* **54:** 4545–4549.

Ann. N.Y. Acad. Sci. ISSN 0077-8923

ANNALS OF THE NEW YORK ACADEMY OF SCIENCES
Issue: *Advances Against Aspergillosis*

The impact of azole resistance on aspergillosis guidelines

Sarah P. Georgiadou and Dimitrios P. Kontoyiannis

Department of Infectious Diseases, Infection Control and Employee Health, The University of Texas MD Anderson Cancer Center, Houston, Texas

Address for correspondence: Dr. Dimitrios P. Kontoyiannis, Department of Infectious Diseases, Infection Control and Employee Health, Unit 402, The University of Texas MD Anderson Cancer Center, 1515 Holcombe Boulevard, Houston, TX 77030. dkontoyi@mdanderson.org

Azole resistance in *Aspergillus* species may be on the rise, with significant potential implications for the management of invasive aspergillosis. The main mechanism of azole resistance in *Aspergillus fumigatus* is via alterations of the target enzyme CYP51A. Such azole resistance is either primary or secondary (in the setting of prior azole exposure) and can be derived either from single or multiple mutations. Irrespective of the amino acid substitution type in CYP51A, azole-resistant *Aspergillus* isolates are always itraconazole resistant. There is significant variability among studies and centers in the prevalence of azole resistance, and this is a multifactorial issue. Nevertheless, the exact frequency of azole resistance is unknown, in part because of the low culturability of the fungus in patients with aspergillosis. This work aims to provide an overview of the current knowledge in *Aspergillus* azole resistance and raises questions for future research and practical implications in the management of aspergillosis.

Keywords: azoles; *Aspergillus*, resistance; guidelines

Introduction

Antifungal drug resistance is an emerging but likely underestimated public health concern given the lack of new antifungal agents in the pipeline.[1] The potential implications of azole resistance in the management of aspergillosis could be significant; voriconazole (VRC), a second-generation triazole and the recommended first-line antifungal agent for all types of invasive aspergillosis (IA),[2] would not be the optimal drug for azole-resistant IA, as its use would probably result in high failure rates. Furthermore, as the use of broad spectrum oral triazoles, such as itraconazole (ITC), VRC, and posaconazole (POSA) (given commonly as primary or secondary prophylaxis), is widespread, physicians would have to resort to logistically inconvenient, expensive, and toxic parenteral drugs in patients with azole-resistant IA. In this review, we attempt to describe the magnitude of the problem regarding epidemiology, mechanisms of *Aspergillus* azole resistance, and practical implications in the clinical management of azole-resistant IA, and we raise questions for future research.

Proposed nomenclature and mechanisms of aspergillosis azole resistance

Undoubtedly, proposed standardized nomenclature regarding *Aspergillus in vitro* resistance for clinically licensed azoles, based on both phenotypic and genotypic definitions, has been a significant advance in the field.[3] For example, a pan-azole-resistant *Aspergillus* isolate is characterized by high minimum inhibitory concentrations (MICs) for all azoles.[3] A multiazole-resistant *Aspergillus* isolate is an isolate whose MICs are in the resistance range for more than one but not all mold-active azoles.[3] An *Aspergillus* isolate resistant to an individual azole (e.g., ITC) is one that has MIC within the resistance range for only one azole.[3] Importantly, there is no definition for dose-dependent azole resistance for *Aspergillus* species as has been described for *Candida* spp. [4]

Azole resistance can be viewed either as primary (intrinsic) or secondary resistance (in the setting of prior azole exposure)[5] and can be derived either from single or multiple mutations.[6] The main mechanism of azole resistance in *Aspergillus fumigatus*

doi: 10.1111/j.1749-6632.2012.06795.x
Ann. N.Y. Acad. Sci. 1272 (2012) 15–22 © 2012 New York Academy of Sciences.

is via alterations, either point mutations or overexpression, of the target enzyme CYP51A.[7–12] Of interest, although there has been no report of azole resistance in IA caused by *A. flavus*, investigators recently described a different mutation of the CYP51C gene for *A. flavus* than the mutations in the CYP51A gene of *A. fumigatus*.[13] Azole resistance due to decreased azole accumulation in cases of increased activity of efflux pumps (ABC, major facilitator superfamily transporters) has been also demonstrated, mainly in laboratory strains;[14] however, the clinical significance of the latter resistance mechanism has yet to be proven. Nevertheless, several azole-resistant *A. fumigatus* isolates do not have any alterations in the target enzyme or in efflux pump activity, suggesting that other uncharacterized mechanisms (e.g., reduced azole uptake) might underlie such resistance.[15]

Investigators have shown in animal models that azole-resistant *A. fumigatus* strains retain their fitness and infected animals have high fungal burden, suboptimal protection by azoles, and increased mortality.[16] In contrast, *A. fumigatus* resistance to other antifungals such as echinocandins or amphotericin B is rare, probably reflecting a fitness loss of those isolates.[17–19] Furthermore, there are specific non-*fumigatus Aspergillus* species, such as *A. terreus*[20] and *A. nidulans*,[21] that have tolerance or even intrinsic resistance to polyenes. It is important to realize that some of the azole-resistant *A. fumigatus* isolates might be "cryptic" rare strains that are erroneously identified as *A. fumigatus* in routine clinical laboratory testing. Some such isolates have been shown to belong to *A. ustus*,[22] *A. lentulus*,[23] and *A. calidoustus*.[24] Similarly, cryptic species were recently found in azole-resistant clinical isolates of the *A. niger* complex.[25] ITC resistance was high, but azole cross-resistance was unusual and the mechanism of resistance remains obscure.[25]

The association between genotype and phenotype in azole resistance for *A. fumigatus* is complex in the setting of the target enzyme CYP51A alterations. Cross-resistance between azoles is related to the specific amino acid substitution.[3] Importantly, the majority of studies have shown that irrespective of the substitution type, these isolates are always ITC resistant.[26] Therefore, ITC resistance may serve as a practical screening marker for azole-resistant *Aspergillus* isolates that would ultimately need further evaluation regarding the genotypic alteration and the degree of cross-resistance or tolerance to other triazoles. Of interest, VRC-only resistance (i.e., without cross-resistance to ITC/POSA) has been reported in the literature, albeit without a causative mechanism of resistance.[27,28]

Issues with current *Aspergillus* azole resistance *in vitro* testing

Optimal *in vitro* testing for azole resistance in *Aspergillus* spp. remains a challenge. Of note, current *in vitro* susceptibility testing is culture based and this is a major limitation in view of the suboptimal culturability of the organism. For example, the diagnostic culture yield of bronchoalveolar lavage in patients with IA is no more than 30%.[29] Furthermore, cultures might take up to five to seven days for completion, so clinicians do not have real time information for the presence of azole resistance.[30] The *in vitro* conditions employed in susceptibility testing are also artificial. Specifically, the methods used in the setting of high inoculum of conidiae in an aerobic, high-glucose environment do not simulate *in vivo* growth of the fungus, which is characterized by a low inoculum, hyphal growth, poor nutrient substrates, semianaerobic environment, and possibly biofilm formation.[31] For example, *Aspergillus* isolates that reside in cavities could develop azole tolerances in the setting of hypoxia.[32] To complicate things further, in a recent study researchers observed diverse microevolutionary alterations in *Aspergillus* isolates residing in cavitary lesions, revealing differences in azole susceptibility, mechanisms of resistance, and genetic type even within the same aspergilloma lesion.[33] In addition, *in vitro* susceptibility tests of *Aspergillus* spp. do not address the heteroresistance phenomenon or fitness cost that renders resistant strains less pathogenic. Most if not all routine microbiological laboratories do not routinely screen for cryptic *Aspergillus* spp., such as *A. lentulus*. Finally, current *in vitro* susceptibility tests do not address the correlation of the genotype derived from either a single nucleotide alteration or multiple mutations of CYP51 with *in vivo* azole resistance and azole cross-resistance.

Epidemiology of azole-resistant aspergillosis

During the past five years, there has been an increasing awareness that azole resistance in *A. fumigatus* might be on the rise, based mainly on the observations of two European medical centers that care for different types of patient populations: Manchester Reference Center (UK), where researchers reported the presence of azole resistance mainly in azole preexposed patients with chronic pulmonary aspergillosis in the setting of underlying structural lung disease,[34,35] and the Nijmegen Medical Center (the Netherlands), where researchers reported the presence of azole resistance in patients with intrinsic *Aspergillus* resistance (no prior azole use), even in patients without underlying chronic pulmonary disease.[11,36,37] Epidemiological exposures such as agricultural fungicides have been also implicated as considerable contributors to azole resistance in the latter center.[38–42] Azole resistance seems to be a relatively recent phenomenon in several other European medical centers as well,[43,44] and its frequency (of both azole and multiazole resistance) seems to be increasing in *A. fumigatus*.[35]

In a recent study, investigators tried to address the regional magnitude of azole resistance;[36] seven university medical centers from the Netherlands participated over a 20-month period, during which all *Aspergillus* isolates were routinely tested for azole resistance in ITC plates (4 mg/L) and were incubated at 35–37 °C. *Aspergillus* isolates were identified by routine macroscopic and microscopic methods. For azole-resistant *Aspergillus* isolates identified after plating in ITC plates, broth microdilution MICs for ITC, VRC, and POSA were then performed, as well as sequencing of the target enzyme CYP51A gene. In this particular study, resistance prevalence was 5.3% per patient and 4.5% per sample; nevertheless, the prevalence varied widely (0.8–95%) between the participant medical centers, even at centers in close geographical proximity to each other. Over 80% of the *Aspergillus* spp. isolates were derived from respiratory samples, and not many cryptic *A. fumigatus* spp. were found (only 2% of isolates). The dominant mechanism of resistance was a TR/L98H mutation in the CYP51A gene that had been previously described by the same investigators.[35,45] Eighty percent of those isolates were cross-resistant to VRC; notably, the degree of resistance to VRC varied widely from 1 mg/L to 16 mg/L. In addition, 17% of ITC-resistant *A. fumigatus* isolates were also resistant to POSA. Interestingly, 64% of the patients were azole naive at the time the resistant *Aspergillus* isolate was cultured. There was no difference in the frequency of underlying diseases between patients with azole-resistant compared with susceptible *A. fumigatus* isolates; nevertheless there was a trend for patients with hematologic malignancies to harbor resistant *Aspergillus* isolates. Overall, the outcome for patients with azole-resistant IA was poor, with over 88% crude mortality,[36] although IA-attributable mortality was not estimated.

Furthermore, attempts have been made recently to evaluate the global distribution of azole resistance in *Aspergillus* species. A global network (SCARE Network) has been created in 20 participating countries from four different continents, based on a culture-based prospective surveillance.[46] Of the 2,449 unselected *A. fumigatus* isolates screened on ITC plates, 51 isolates (5.3%) were found to be ITC resistant. Similar to findings in the Netherlands study,[36] significant prevalence variability was described among the different centers (0–4.2%). Azole resistance was demonstrated in many but not all countries. In this study, 11 of the 51 (22%) ITC-resistant *A. fumigatus*-like isolates were identified as "sibling" species; of these 11 isolates, 5 (45%) were *A. lentulus*. Moreover, in agreement with most of the European studies, the main mechanism of resistance was due to the TR/L98H mutation, and 82% of the patients were azole naive. In contrast, in a recent multicenter study in which 1789 *Aspergillus* isolates from 63 medical centers worldwide were tested using an epidemiological cut-off value, *in vitro* VRC resistance was low (less than 2%) and there was no consistent trend toward decreased susceptibility for any broad spectrum triazole during a 9-year period (2001–2009).[47] Of interest, geographical niches of presumably nonwild-type *Aspergillus* isolates having high MICs to azoles were identified outside Europe, especially in China.[47,48]

Prevalence of aspergillosis azole resistance: a multifactorial issue

Undoubtedly, the tremendous variability of azole resistance prevalence among previous studies is multifactorial (Table 1): patient referral patterns, selection versus random testing of specimens, time of the study (older versus recent studies), number of isolates tested (studies examining more versus fewer than 100 isolates), differences in the screening methods (e.g., plate-based versus broth-based), work-up to exclude cryptic *Aspergillus* spp. (e.g., *A. lentulus*), differences in the definitions of *in vitro* resistance, and the testing methods or inocula

Table 1. Variability of azole resistance prevalence in *Aspergillus* isolates among medical centers and published studies: a multifactorial issue

---

Parameters

Patient referral patterns (acute versus chronic aspergillosis)

Selected versus random specimens

Time factors (recent versus older series)

Small collections of isolates (e.g., <100 isolates)

Differences in screening methods (microdilution versus plate method)

Exclusion of cryptic *Aspergillus* species

Rates using different denominators (patient versus isolates)

Different susceptibility methods and inocula

Different breakpoints[a] and definitions of resistance

Differences in geographic location of the center

---

[a]In the absence of clinical breakpoints, epidemiological cutoff values have been established as a means of distinguishing wild-type isolates of *Aspergillus* spp. by the Clinical and Laboratory Standards Institute (CLSI).[65]

account for this variability. Importantly, it is possible that the prevalence of *A. fumigatus* azole resistance is underestimated if culture-independent, molecular methods are employed. In a recent study, in which investigators used an ultrasensitive PCR amplification of the CYP51A gene in sputum specimens of both patients with different aspergillosis syndromes and healthy volunteers, it was found that PCR was significantly more sensitive than culture-based methods.[34] Specifically, PCR was positive for *Aspergillus* spp. in 79% and 71% of patients with allergic and chronic pulmonary aspergillosis, compared with 0% and 16% by culture, respectively. In culture-negative PCR-positive cases, resistance mutations were found in 55% of tested samples, which raises the question whether those isolates have a fitness cost and do not grow easily in cultures. Resistance markers were seen even in patients without prior azole exposure, as well as in patients with concurrent azole therapy and adequate serum azole levels, although the data were too limited to draw valid conclusions. Nevertheless, this study raises the question whether PCR is a better way to detect both *Aspergillus* spp. and azole resistance markers, at least

in aspergillosis syndromes that are associated with low inoculum disease.

However, the controversy regarding the impact of *in vitro* azole resistance in *A. fumigatus* is enhanced by the fact that clinical failure in IA syndromes is multifactorial and associated not only with the fungal isolate and its possible biological fitness, but also with the pharmacology of the antifungal agent used and with a variety of host factors (Table 2). For example, the mode of action of antifungal drugs (fungistatic versus fungicidal) and their pharmacokinetics (absorption, tissue distribution, metabolism, drug–drug interactions) are significant factors in prognosis.[49] In the big picture, host factors continue to be the most important determinants of clinical failure of azoles in IA, due to poor immune status, site of infection, presence of tissue sequestration, severity of infection, and patient noncompliance with the drug regimen (especially in chronic IA syndromes that require several months of antifungal treatment).

Unresolved issues of azole resistance for the clinician

Whether *in vitro* azole resistance is an *independent* risk factor for clinical failure remains uncertain. Notably, previous studies regarding azole-resistant IA usually did not report on the time of onset of antifungals from the IA diagnosis, the activity of the underlying disease, whether antifungals were given intermittently or continuously or whether antifungals were given as monotherapy or in combination treatment, and the effect of mixed or subsequent infections.[11,34–36] This controversy is further enhanced by the fact that the recently described transplant-associated infection surveillance network (TRANSNET) did not report any mortality difference between patients with IA caused by either VRC-resistant or VRC-sensitive *Aspergillus* isolates.[50] The effect of pharmacokinetic parameters such as the area under the curve (AUC) and AUC/MIC of each azole, and the importance of fluctuations of azole serum levels in relationship to the development and outcome of IA caused by azole-resistant *Aspergillus* isolates, are obscure. Interestingly, although most patients with chronic aspergillosis symptoms fail to improve and remain stable, not all actually progress.[35] This observation suggests that in at least that patient population, structural underlying lung disease,

Table 2. Azole failure in invasive aspergillosis: a multifactorial issue

Host	Azole	Pathogen
Prior exposure to azoles	Intermittent versus constant exposure	Species-specific variability
Cavitary lung lesions[a]	Decreased AUC/MIC	Isolate-specific variability
Compliance	Volume of distribution	Fungal burden
Poor systemic and infection site–specific immune responses[a]	Lipophilicity	
Delayed diagnosis[a]	Sequence of azole use (e.g., FLU → POSA, ITC → POSA)	

[a]Predisposing factors for high fungal burden.
AUC, area under the curve; FLU, fluconazole; ITC, itraconazole; MIC, minimal inhibitory concentration; POSA, posaconazole.

fitness loss of resistant *Aspergillus* strains, and azole resistance may induce competing scenarios. Similarly, the suitability of using another azole sequentially and the optimal sequence of different classes of antifungal agents (either alone or in combination), as well as the "threshold" of resistance in order to start a non-VRC-containing regimen in patients with IA in each center, are still unknown. Finally, there are no studies regarding the potential effect on *Aspergillus* galactomannan kinetics in azole-resistant *Aspergillus* isolates under various treatments.

Future perspective and practical considerations

Undoubtedly, there is an urgent need for continuation of global surveillance of both clinical and environmental *Aspergillus* spp. isolates that should be based on standardized nomenclature and laboratory detection methods. A practical way to screen for azole resistance could be based on plating on ITC plates, as well as on the regular performance of thermotolerance tests to look for *A. lentulus*.[23] Furthermore, as azole resistance has been described even in patients who received antifungal treatment for less than one month,[36] it might be reasonable to screen for azole resistance all patients with culture-documented IA who are to proceed to hematopoietic stem cell transplantation or reinduction cytotoxic chemotherapy for acute leukemia, in order to estimate whether different therapeutic antifungal strategies should be adopted. We believe that in centers where the prevalence of azole resistance is known to be high (e.g., over 5%), initial treatment with different classes of antifungal agents such as an echinocandin with liposomal amphotericin B (each alone or in combination) might be prudent, although this is yet to be demonstrated. Of note, previous *in vitro* studies on the synergistic effect of combination of VRC with an echinocandin against VRC-resistant *Aspegillus* spp. showed conflicting results.[51–53] In addition, there is an urgent necessity to expand on research in PCR detection methods both in bronchoalveolar lavage and tissues, in order to employ more sensitive tests than cultures for IA diagnosis, as well as azole-resistance mutation(s) identification.[54,55] Finally, as *Aspergillus* galactomannan (GM) is increasingly considered a surrogate marker for response,[56,57] a rising GM titer in a patient with documented IA who is on triazole-based therapy might be indicative of azole-resistant IA, especially if there is documentation of adequate serum triazole levels. Further preclinical and clinical research is needed on the role of GM as a marker of azole-resistant aspergillosis.

However, as we enter the era of genomic medicine and the cost of DNA sequencing is becoming affordable (sequencing the whole human genome currently costs about $1,000), we wonder whether deep-sequencing on a large scale of *Aspergillus* isolates, and not only of the CYP51A gene, might be the answer needed to address the complex issues of compensatory mutations, multiple targets, etc. This approach proved to be valid when large-scale phenotypic and genotypic data were correlated with patient-level clinical data leading to the establishment of compensatory mutation roles in the management of human immunodeficiency virus infection.[58] Disturbingly, autopsy rates are very low in cancer care hospitals in the Western world (e.g., at MD Anderson Cancer Center the current autopsy rate is less than 5%); therefore, we might not be aware of the true magnitude of this problem. However, the little available data we have so far suggest that azoles might have had a major positive effect in the prevalence of IA at autopsy; for example, data from our tertiary care cancer suggest that the incidence of IA in the era of widespread use of *Aspergillus*-active triazoles has gradually decreased (Kontoyiannis, submitted).

Nonetheless, it should not be extrapolated that azole resistance in *Aspergillus* will eventually result in a worldwide pandemic. The roles of the unique host and geographic niches of resistant opportunistic fungal infections are reminders of the fact that fungal resistance is a complex multifaceted issue. For example, fungal infections with *Candida krusei*[59] or *Fusarium solani*[60] did not spread beyond the niches of leukemia and hematopoietic stem cell transplantation patient populations. The same observation applies for *Scedosporium prolificans* infections that have been both geographically restricted (e.g., Australia and Spain) and typically reported in patients with underlying cystic fibrosis.[61] Likewise, *A. nidulans* is a typical pathogen only in patients with chronic granulomatous disease,[62] and *A. terreus* has been described mainly in isolated centers of Austria[63] and Texas.[64]

Conclusions

Resistance of *Aspergillus* spp. to azoles might be on the rise; nevertheless, inferences are drawn from incomplete observational data, and it is rather early to consider that azole resistance is becoming a menacing worldwide phenomenon. A better understanding of the epidemiology and the biologic plausibility of resistance would assist in the development of better diagnostic strategies (probably PCR-based assays) for both prevention and treatment. Currently, we need better field studies but not necessarily modification of guidelines.

Acknowledgments

D.P.K. acknowledges the Frances King Black Endowed Professorship for Cancer Research.

Conflicts of interest

DPK has received research support and honoraria from Pfizer; Astellas Pharma U.S.; Gilead Sciences, Inc.; and Merck and Co., Inc. SPG reports no conflicts.

References

1. Ananda-Rajah, M.R., M.A. Slavin & K.T. Thursky. 2012. The case for antifungal stewardship. *Curr. Opin. Infect. Dis.* **25:** 107–115.
2. Walsh, T.J., E.J. Anaissie, D.W. Denning, *et al.* 2008. Treatment of aspergillosis: clinical practice guidelines of the Infectious Diseases Society of America. *Clin. Infect. Dis.* **46:** 327–360.
3. Verweij, P.E., S.J. Howard, W.J. Melchers, *et al.* 2009. Azole-resistance in Aspergillus: proposed nomenclature and breakpoints. *Drug Resist. Update* **12:** 141–147.
4. Pfaller, M.A., D.J. Diekema, J.H. Rex, *et al.* 2006. Correlation of MIC with outcome for Candida species tested against voriconazole: analysis and proposal for interpretive breakpoints. *J. Clin. Microbiol.* **44:** 819–826.
5. Bowyer, P., C.B. Moore, R. Rautemaa, *et al.* 2011. Azole antifungal resistance today: focus on Aspergillus. *Curr. Infect. Dis. Rep.* **13:** 485–491.
6. Camps, S.M., J.W. van der Linden, Y. Li, *et al.* 2012. Rapid induction of multiple resistance mechanisms in Aspergillus fumigatus during azole therapy: a case study and review of the literature. *Antimicrob. Agents Chemother.* **56:** 10–16.
7. Chen, J., H. Li, R. Li, *et al.* 2005. Mutations in the cyp51A gene and susceptibility to itraconazole in Aspergillus fumigatus serially isolated from a patient with lung aspergilloma. *J. Antimicrob. Chemother* **55:** 31–37.
8. Howard, S.J., I. Webster, C.B. Moore, *et al.* 2006. Multi-azole resistance in Aspergillus fumigatus. *Int. J. Antimicrob. Agents* **28:** 450–453.
9. Mellado, E., G. Garcia-Effron, L. Alcazar-Fuoli, *et al.* 2004. Substitutions at methionine 220 in the 14alpha-sterol demethylase (Cyp51A) of Aspergillus fumigatus are responsible for resistance in vitro to azole antifungal drugs. *Antimicrob. Agents Chemother.* **48:** 2747–2750.
10. Mellado, E., G. Garcia-Effron, L. Alcazar-Fuoli, *et al.* 2007. A new Aspergillus fumigatus resistance mechanism conferring in vitro cross-resistance to azole antifungals involves a combination of cyp51A alterations. *Antimicrob. Agents Chemother.* **51:** 1897–1904.
11. Snelders, E., H.A. van der Lee, J. Kuijpers, *et al.* 2008. Emergence of azole resistance in Aspergillus fumigatus and spread of a single resistance mechanism. *PLoS Med.* **5:** e219.
12. Verweij, P.E., E. Mellado & W.J. Melchers. 2007. Multiple-triazole-resistant aspergillosis. *N. Engl. J. Med.* **356:** 1481–1483.
13. Liu, W., Y. Sun, W. Chen, *et al.* 2012. T778G mutation in cyp51C gene confers voriconazole-resistance in Aspergillus flavus causing aspergillosis. *Antimicrob. Agents Chemother* **56:** 2598–2603.
14. Slaven, J.W., M.J. Anderson, D. Sanglard, *et al.* 2002. Increased expression of a novel Aspergillus fumigatus ABC

transporter gene, atrF, in the presence of itraconazole in an itraconazole resistant clinical isolate. *Fungal Genet. Biol.* **36:** 199–206.

15. Manavathu, E.K., J. Cutright & P.H. Chandrasekar. 1999. Comparative study of susceptibilities of germinated and ungerminated conidia of Aspergillus fumigatus to various antifungal agents. *J. Clin. Microbiol.* **37:** 858–861.

16. Dannaoui, E., E. Borel, F. Persat, *et al.* 1999. In-vivo itraconazole resistance of Aspergillus fumigatus in systemic murine aspergillosis. EBGA Network. European research group on Biotypes and Genotypes of Aspergillus fumigatus. *J. Med. Microbiol.* **48:** 1087–1093.

17. Chamilos, G. & D.P. Kontoyiannis. 2005. Update on antifungal drug resistance mechanisms of Aspergillus fumigatus. *Drug Resist. Update* **8:** 344–358.

18. Pfaller, M.A., L. Boyken, R.J. Hollis, *et al.* 2009. In vitro susceptibility of clinical isolates of Aspergillus spp. to anidulafungin, caspofungin, and micafungin: a head-to-head comparison using the CLSI M38-A2 broth microdilution method. *J. Clin. Microbiol.* **47:** 3323–3325.

19. Pfaller, M.A., L. Boyken, R.J. Hollis, *et al.* 2010. Wild-type minimum effective concentration distributions and epidemiologic cutoff values for caspofungin and Aspergillus spp. as determined by Clinical and Laboratory Standards Institute broth microdilution methods. *Diagn. Microbiol. Infect. Dis.* **67:** 56–60.

20. Steinbach, W.J., D.K. Benjamin, Jr., D.P. Kontoyiannis, *et al.* 2004. Infections due to Aspergillus terreus: a multicenter retrospective analysis of 83 cases. *Clin. Infect. Dis.* **39:** 192–198.

21. Kontoyiannis, D.P., R.E. Lewis, G.S. May, *et al.* 2002. Aspergillus nidulans is frequently resistant to amphotericin B. *Mycoses* **45:** 406–407.

22. Varga, J., J. Houbraken, H.A. Van Der Lee, *et al.* 2008. Aspergillus calidoustus sp. nov., causative agent of human infections previously assigned to Aspergillus ustus. *Eukaryot. Cell* **7:** 630–638.

23. Balajee, S.A., J.L. Gribskov, E. Hanley, *et al.* 2005. Aspergillus lentulus sp. nov., a new sibling species of A. fumigatus. *Eukaryot. Cell* **4:** 625–632.

24. Egli, A., J. Fuller, A. Humar, *et al.* 2012. Emergence of Aspergillus calidoustus infection in the era of posttransplantation azole prophylaxis. *Transplantation* **94:** 403–410.

25. Howard, S.J., E. Harrison, P. Bowyer, *et al.* 2011. Cryptic species and azole resistance in the Aspergillus niger complex. *Antimicrob. Agents Chemother.* **55:** 4802–4809.

26. Perlin, D.S. 2009. Antifungal drug resistance: do molecular methods provide a way forward? *Curr. Opin. Infect. Dis.* **22:** 568–573.

27. Bueid, A., S.J. Howard, C.B. Moore, *et al.* 2010. Azole antifungal resistance in Aspergillus fumigatus: 2008 and 2009. *J. Antimicrob. Chemother.* **65:** 2116–2118.

28. Kuipers, S., R.J. Bruggemann, R.G. de Sevaux, *et al.* 2011. Failure of posaconazole therapy in a renal transplant patient with invasive aspergillosis due to Aspergillus fumigatus with attenuated susceptibility to posaconazole. *Antimicrob. Agents Chemother.* **55:** 3564–3566.

29. Tarrand, J.J., M. Lichterfeld, I. Warraich, *et al.* 2003. Diagnosis of invasive septate mold infections. A correlation of microbiological culture and histologic or cytologic examination. *Am. J. Clin. Pathol.* **119:** 854–858.

30. Tarrand, J.J., X.Y. Han, D.P. Kontoyiannis, *et al.* 2005. Aspergillus hyphae in infected tissue: evidence of physiologic adaptation and effect on culture recovery. *J. Clin. Microbiol.* **43:** 382–386.

31. Muller, F.M., M. Seidler & A. Beauvais. 2011. Aspergillus fumigatus biofilms in the clinical setting. *Med. Mycol.* **49(Suppl 1):** S96–S100.

32. Willger, S.D., S. Puttikamonkul, K.H. Kim, *et al.* 2008. A sterol-regulatory element binding protein is required for cell polarity, hypoxia adaptation, azole drug resistance, and virulence in Aspergillus fumigatus. *PLoS Pathog.* **4:** e1000200.

33. Howard, S.J., *et al.* 2012. Microevolution of Aspergillus Fumigatus in aspergillomas. In *5th Advances Against Aspergillosis, P-124*, Istanbul, Turkey.

34. Denning, D.W., S. Park, C. Lass-Florl, *et al.* 2011. High-frequency triazole resistance found In nonculturable Aspergillus fumigatus from lungs of patients with chronic fungal disease. *Clin. Infect. Dis.* **52:** 1123–1129.

35. Howard, S.J., D. Cerar, M.J. Anderson, *et al.* 2009. Frequency and evolution of Azole resistance in Aspergillus fumigatus associated with treatment failure. *Emerg. Infect. Dis.* **15:** 1068–1076.

36. van der Linden, J.W., E. Snelders, G.A. Kampinga, *et al.* 2011. Clinical implications of azole resistance in Aspergillus fumigatus, The Netherlands, 2007–2009. *Emerg. Infect. Dis.* **17:** 1846–1854.

37. Verweij, P.E., D.T. Te Dorsthorst, A.J. Rijs, *et al.* 2002. Nationwide survey of in vitro activities of itraconazole and voriconazole against clinical Aspergillus fumigatus isolates cultured between 1945 and 1998. *J. Clin. Microbiol.* **40:** 2648–2650.

38. Snelders, E., S.M. Camps, A. Karawajczyk, *et al.* 2012. Triazole fungicides can induce cross-resistance to medical triazoles in Aspergillus fumigatus. *PLoS One* **7:** e31801.

39. Mortensen, K.L., E. Mellado, C. Lass-Florl, *et al.* 2010. Environmental study of azole-resistant Aspergillus fumigatus and other aspergilli in Austria, Denmark, and Spain. *Antimicrob. Agents Chemother.* **54:** 4545–4549.

40. Snelders, E., R.A. Huis In 't Veld, A.J. Rijs, *et al.* 2009. Possible environmental origin of resistance of Aspergillus fumigatus to medical triazoles. *Appl. Environ. Microbiol.* **75:** 4053–4057.

41. Verweij, P.E., E. Snelders, G.H. Kema, *et al.* 2009. Azole resistance in Aspergillus fumigatus: a side-effect of environmental fungicide use? *Lancet Infect. Dis.* **9:** 789–795.

42. Klaassen, C.H., J.G. Gibbons, N.D. Fedorova, *et al.* 2012. Evidence for genetic differentiation and variable recombination rates among Dutch populations of the opportunistic human pathogen Aspergillus fumigatus. *Mol. Ecol.* **21:** 57–70.

43. Morio, F., G.G. Aubin, I. Danner-Boucher, *et al.* 2012. High prevalence of triazole resistance in Aspergillus fumigatus, especially mediated by TR/L98H, in a French cohort of patients with cystic fibrosis. *J. Antimicrob. Chemother.* **67:** 1870–1873.

44. Pelaez, T., P. Gijon, E. Bunsow, *et al.* 2012. Resistance to voriconazole due to a G448S substitution in Aspergillus fumigatus in a patient with cerebral aspergillosis. *J. Clin. Microbiol.* **50:** 2531–2534.

45. Snelders, E., A. Karawajczyk, G. Schaftenaar, *et al.* 2010. Azole resistance profile of amino acid changes in Aspergillus

fumigatus CYP51A based on protein homology modeling. *Antimicrob. Agents Chemother.* **54:** 2425–2430.

46. Van der Linden, J.W. M. *et al.* 2011. *Prospective International Surveillance of Azole Resistance in Aspergillus fumigatus. SCARE—Network. In* ICAAC, M-490, San Francisco, USA.

47. Pfaller, M., L. Boyken, R. Hollis, *et al.* 2011. Use of epidemiological cutoff values to examine 9-year trends in susceptibility of Aspergillus species to the triazoles. *J. Clin. Microbiol.* **49:** 586–590.

48. Lockhart, S.R., J.P. Frade, K.A. Etienne, *et al.* 2011. Azole resistance in Aspergillus fumigatus isolates from the ARTEMIS global surveillance study is primarily due to the TR/L98H mutation in the cyp51A gene. *Antimicrob. Agents Chemother.* **55:** 4465–4468.

49. Lewis, J.S., 2nd & J.R. Graybill. 2008. Fungicidal versus Fungistatic: what's in a word? *Expert Opin. Pharmacother.* **9:** 927–935.

50. Kontoyiannis, D.P., K.A. Marr, B.J. Park, *et al.* 2010. Prospective surveillance for invasive fungal infections in hematopoietic stem cell transplant recipients, 2001–2006: overview of the Transplant-Associated Infection Surveillance Network (TRANSNET) Database. *Clin. Infect. Dis.* **50:** 1091–1100.

51. Planche, V., S. Ducroz, A. Alanio, *et al.* 2012. In vitro combination of anidulafungin and voriconazole against intrinsically azole-susceptible and -resistant Aspergillus spp. *Antimicrob. Agents Chemother.* **56:** 4500–4503.

52. Jeans, A.R., S.J. Howard, Z. Al-Nakeeb, *et al.* 2012. Combination of voriconazole and anidulafungin for the treatment of triazole resistant Aspergillus fumigatus in an in vitro model of invasive pulmonary aspergillosis. *Antimicrob. Agents Chemother* **56:** 5180–5185.

53. Krishnan-Natesan, S., W. Wu & P.H. Chandrasekar. 2012. In vitro efficacy of the combination of voriconazole and anidulafungin against voriconazole-resistant cyp51A mutants of Aspergillus fumigatus. *Diagn. Microbiol. Infect. Dis.* **73:** 135–137.

54. van der Linden, J.W., E. Snelders, J.P. Arends, *et al.* 2010. Rapid diagnosis of azole-resistant aspergillosis by direct PCR using tissue specimens. *J. Clin. Microbiol.* **48:** 1478–1480.

55. Spiess, B., W. Seifarth, N. Merker, *et al.* 2012. Development of novel PCR assays to detect azole resistance-mediating mutations of the Aspergillus fumigatus cyp51A gene in primary clinical samples from neutropenic patients. *Antimicrob. Agents Chemother.* **56:** 3905–3910.

56. Koo, S., J.M. Bryar, L.R. Baden, *et al.* 2010. Prognostic features of galactomannan antigenemia in galactomannan-positive invasive aspergillosis. *J. Clin. Microbiol.* **48:** 1255–1260.

57. Miceli, M.H., M.L. Grazziutti, G. Woods, *et al.* 2008. Strong correlation between serum aspergillus galactomannan index and outcome of aspergillosis in patients with hematological cancer: clinical and research implications. *Clin. Infect. Dis.* **46:** 1412–1422.

58. Panel on Antiretroviral Guidelines for Adults and Adolescents. Guidelines for the use of antiretroviral agents in HIV-infected adults and adolescents. Department of Health and Human Services. October 14, 2011; 1–167 : Available at: http://www.aidsinfo.nih.gov/ContentFiles/AdultandAdolescentGL.pdf.

59. Miceli, M.H., J.A. Diaz & S.A. Lee. 2011. Emerging opportunistic yeast infections. *Lancet Infect. Dis.* **11:** 142–151.

60. Campo, M., R.E. Lewis & D.P. Kontoyiannis. 2010. Invasive fusariosis in patients with hematologic malignancies at a cancer center: 1998–2009. *J. Infect.* **60:** 331–337.

61. Rodriguez-Tudela, J.L., J. Berenguer, J. Guarro, *et al.* 2009. Epidemiology and outcome of Scedosporium prolificans infection, a review of 162 cases. *Med. Mycol.* **47:** 359–370.

62. Blumental, S., R. Mouy, N. Mahlaoui, *et al.* 2011. Invasive mold infections in chronic granulomatous disease: a 25-year retrospective survey. *Clin. Infect. Dis.* **53:** e159–169.

63. Lass-Florl, C., K. Griff, A. Mayr, *et al.* 2005. Epidemiology and outcome of infections due to Aspergillus terreus: 10-year single centre experience. *Br. J. Haematol.* **131:** 201–207.

64. Hachem, R.Y., D.P. Kontoyiannis, M.R. Boktour, *et al.* 2004. Aspergillus terreus: an emerging amphotericin B-resistant opportunistic mold in patients with hematologic malignancies. *Cancer* **101:** 1594–1600.

65. Espinel-Ingroff, A., D.J. Diekema, A. Fothergill, *et al.* 2010. Wild-type MIC distributions and epidemiological cutoff values for the triazoles and six Aspergillus spp. for the CLSI broth microdilution method (M38-A2 document). *J. Clin. Microbiol.* **48:** 3251–3257.

Ann. N.Y. Acad. Sci. ISSN 0077-8923

ANNALS OF THE NEW YORK ACADEMY OF SCIENCES

Issue: *Advances Against Aspergillosis*

Risk stratification for invasive aspergillosis in immunocompromised patients

Raoul Herbrecht,[1] Pierre Bories,[1] Jean-Charles Moulin,[1] Marie-Pierre Ledoux,[1] and Valérie Letscher-Bru[2]

[1]Department of Oncology and Hematology, Hôpital de Hautepierre, Strasbourg, France. [2]Institut de Parasitologie et de Pathologie Tropicale, Strasbourg, France

Address for correspondence: Raoul Herbrecht, Department of Oncology and Hematology, Hôpital de Hautepierre, Hôpitaux Universitaires de Strasbourg, 67098 Strasbourg, France. raoul.herbrecht@chru-strasbourg.fr

Severe prolonged neutropenia, allogeneic hematopoietic stem cell or solid-organ transplantation, corticosteroids or other T cell suppressive agents, and other severe immunosuppressive factors have for many years been considered to predispose patients to invasive aspergillosis. Other conditions such as impaired innate immunity, diabetes, renal impairment, progression of the underlying malignancy, prior respiratory disease, and nosocomial or environmental exposure to fungal spores or climatic factors have recently been considered additional risk factors of invasive aspergillosis. The multiplicity of risk factors as well as the obvious synergy between them renders risk stratification difficult. An international, large-scale, multicenter, epidemiological study is necessary to develop a risk score.

Keywords: invasive aspergillosis; leukemia; neutropenia; hematopoietic stem cell transplantation; risk factors

Introduction

In spite of more effective prophylaxis strategies, invasive aspergillosis remains the most common invasive fungal disease in allogeneic hematopoietic stem cell transplant recipients and in patients receiving induction or consolidation chemotherapy for acute leukemia.[1,2] With the development of new immunosuppressive therapy such as T cell inhibitors or monoclonal antibodies, patients with other hematological malignancies are now also at risk of invasive aspergillosis.[3,4] Among solid-organ transplant recipients, invasive aspergillosis is the second most common invasive fungal disease following invasive candidiasis.[5]

Invasive aspergillosis is still difficult to diagnose and to treat. Direct (microscopy, culture, histopathology) and indirect (*Aspergillus* galactomannan, β-D-glucan, PCR) confirmatory tests lack sensitivity. Therefore precise knowledge of the host factors predisposing to this infection is critical to suspect the diagnosis and to initiate early therapy before mycological confirmation can be obtained.

Incidence of invasive aspergillosis

Many studies have assessed the incidence of invasive aspergillosis in various groups of patients at risk of opportunistic infections. Table 1 lists the most frequent of these predisposing conditions and the range of incidence reported in the literature.[2,6–32] Incidence rates refer to the main underlying conditions and not to all the other factors associated with either a risk of colonization by the fungi or with a risk of tissue invasion by the pathogens. The impact of the combination of host group, risk of colonization and of the various conditions impairing the host response cannot appropriately be deduced from this table to stratify the risk.

How can we stratify the risk?

Identification of the patient population at risk for invasive aspergillosis may also be approached from the characteristics of patients included in recent major multicenter clinical trials conducted on invasive aspergillosis (Fig. 1).[33,34] Hematological malignancies and hematopoietic stem cell transplantations were the two most frequent underlying conditions,

doi: 10.1111/j.1749-6632.2012.06829.x

Table 1. Identified main underlying conditions predisposing to invasive aspergillosis, reported incidence, and specific patient and treatment-related risk factors[2,6–32]

Underlying conditions	Incidence (%)	Identified specific patient and treatment-related risk factors
Transplantation		
Allogeneic hematopoietic stem cells	2.7–23	Delayed neutrophil engraftment, secondary neutropenia, lymphocytopenia, monocytopenia, cord blood, T cell-depleted or CD34-selected stem cell products, unrelated or mismatched donor graft, acute or chronic graft versus host disease, corticosteroids, CMV disease, respiratory virus infections, renal failure, reduced-intensity conditioning regimen, purine analogs or monoclonal antibodies, history of invasive aspergillosis, iron overload, advanced age, donor toll-like receptor polymorphism
Autologous hematopoietic stem cells	0.5–6[a]	Neutropenia, purine analogs or monoclonal antibodies, lymphoproliferative malignancy as indication for transplantation
Lung or heart-lung	3–26	Airways colonization, single lung transplant, cystic fibrosis as indication for transplantation, rejection and increased immunosuppression, obliterative bronchitis, lymphocytopenia, corticosteroids, CMV disease, renal dysfunction, hypogammaglobulinemia, advanced age
Heart	0.4–15	Airways colonization, reoperation, Lymphocytopenia, corticosteroids, CMV disease, post-transplant hemodialysis, advanced age
Liver	0.7–10	Retransplantation, smoking, CMV disease, dialysis requirement, fulminant hepatic failure as indication for transplantation
Pancreas	1.1–2.9	None identified
Kidney	0.2–1	Graft failure requiring prolonged hemodialysis, lymphocytopenia, corticosteroids, sirolimus $\pm$ mycophenolate mofetil, advanced age
Small bowell	0–11	Lymphocytopenia, corticosteroids, CMV disease, renal dysfunction, advanced age
Malignancies		
Acute myeloid leukemia	5–24[b]	Neutropenia, monocytopenia, purine analogs or monoclonal antibodies, advanced age, iron overload, influenza H1N1 virus infection, lack of response to induction chemotherapy
Acute lymphoblastic leukemia	3.8	Lymphocytopenia, corticosteroids, advanced age
Multiple myeloma	2–3	Neutropenia, corticosteroids, advanced age
Non-Hodgkin's lymphoma	0.8	Corticosteroids, purine analogs or monoclonal antibodies, advanced age
Hodgkin's disease	0.4	None identified
Lung cancer	2.6	Stage IV disease, corticosteroids
Other		
AIDS	0–12	<50 CD4$^+$ cells/μL, neutropenia, corticosteroids
Chronic granulomatous disease	20–40[c]	None identified
Burns	1–7	Body surface area burn, full thickness burn, length of hospital stay, older age, inhalation injuries

Continued

Table 1. *Continued*

Underlying conditions	Incidence (%)	Identified specific patient and treatment-related risk factors
Chronic obstructive pulmonary disease with acute exacerbation	1.9	Systemic or inhaled corticosteroids, admission to the intensive-care unit, chronic heart failure, antibiotic treatment received in the three months prior to admission, airways colonization
Systemic lupus erythematous	0.5–2.1	Corticosteroids, other immunosuppressive drugs
Liver failure[d]	5.4	Multiple antibiotic use, frequent invasive procedures
Severe combined immunodeficiency	3.5	None identified

[a]Lower after transplantation of peripheral stem cells compared to bone marrow stem cells.
[b]Lower in acute lymphoblastic leukemia compared to acute myeloblastic leukemia.
[c]Lowered with itraconazole prophylaxis.
[d]Hepatitis B virus-related liver failure.

representing nearly 90% of primary predisposing factors of the host. Solid-organ transplantation, AIDS, and corticosteroid treatment each represented only 3–4% of primary predisposing conditions. Two further open-label studies assessing the efficacy of caspofungin were performed exclusively in patients with a hematological malignancy and in recipients of allogeneic stem cell transplantation.[35,36]

This approach of defining the patient population has certain limitations. In order to exhibit homogeneous patient populations in clinical trials, strict definition criteria are to be applied, with criteria usually based on or adapted from the European Organization for Research and Treatment of Cancer and the Mycosis Study Group (EORTC/MSG).[37] In general, the characteristics of patients included in the clinical trials reflect the inclusion and exclusion criteria, the consensus definition criteria if any exist, as well as the predominant population of patients admitted in the centers selected for the trial.

The real-life clinical setting, however, differs, as a significant proportion of patients seen in our daily practice would not qualify for inclusion in therapeutic clinical trials. In our experience, 22% of subjects in a cohort of 195 patients with documented invasive aspergillosis (probable or proven disease according to the EORTC/MSG definition) would not have been eligible for entry into the two largest clinical trials, notably the voriconazole versus amphotericin B trial

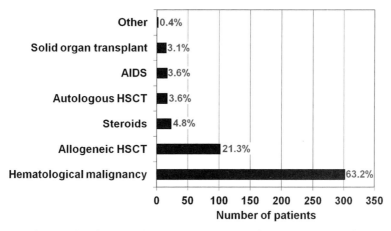

Figure 1. Distribution of primary host factors predisposing to invasive aspergillosis in the voriconazole trial and in the Ambiload trial (total number of patients = 478).[33,34]

and the Ambiload trial.[33,34,38] The main reasons for noneligibility would have been inappropriate renal function, the need for mechanical ventilation, or short life expectancy due to end-stage hematological malignancies (unpublished data). These noneligible patients generally exhibit more severe underlying conditions as well as concomitant single or multiple organ failure, thus creating a different environment for the development of invasive aspergillosis. Due to these comorbidities, these patients also display a poorer prognosis.

Another limitation in qualifying for entry into clinical trials is the lack of host factors, as defined by the EORTC/MSG criteria.[37] While these criteria were designed to help select a homogeneous patient population in clinical trials, they exclude many patients with less typical predisposing factors. As stated in the consensus definition paper, host factor is not synonymous with risk factor. Thus, the lack of host factor may be an exclusion criterion for clinical trials but should not be a reason for deciding to not initiate treatment if clinical, radiological, or mycological data suggest invasive aspergillosis. As an example,

in our center, the median duration of neutropenia prior to the first signs of invasive aspergillosis was 16 days in 91 patients with acute myeloblastic leukemia and invasive aspergillosis (unpublished data). However, 36 (40%) patients had neutropenia for less than 10 days before development of the fungal disease and would therefore not have qualified for the definition of possible, probable or proven invasive aspergillosis according to the EORTC/MSG criteria.[37] Nevertheless, the group of patients with a short duration of neutropenia prior to the first signs more frequently suffered from renal impairment, prior respiratory disease, or diabetes. These data strongly suggest that two or more risk factors may precipitate the occurrence of invasive aspergillosis in patients who do not meet the consensus criteria for host factors. This necessitates greater flexibility in these criteria, while seeking to consider a combination of risk factors in the consensus definitions. In an attempt to stratify the risk factors for invasive *Aspergillus* disease, we must distinguish patient-related factors and environmental factors (Fig. 2).

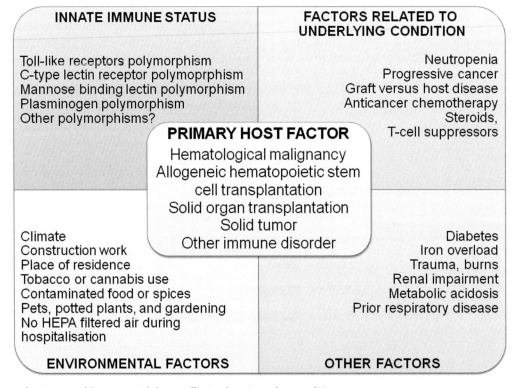

Figure 2. Diagram of the various risk factors affecting the primary host condition.

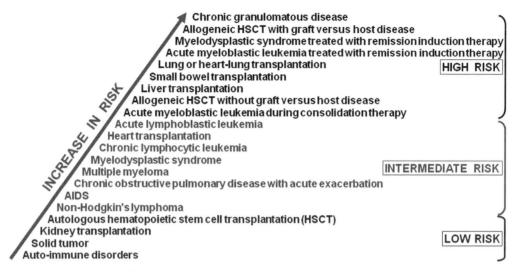

Figure 3. Risk of invasive aspergillosis based on the primary host factor.

Patient-related risk factors

Patient-related risk factors primarily concern the underlying condition, such as allogeneic hematopoietic stem cell transplantation, solid-organ transplantation, prolonged neutropenia occurring in a patient with hematological malignancy or any type of other significant immune disorder (Fig. 3). The treatment given for the underlying disease may be as central as the disease itself. Corticosteroids, T cell suppressors like calcineurin inhibitors (cyclosporine, tacrolimus, sirolimus), monoclonal antibodies with anti-T cell activity (e.g., alemtuzumab), antithymocytes globulins, anti-TNF-α monoclonal antibodies (e.g., infliximab or adalimumab) or other TNF-α inhibitors (etanercept), purine analogs (e.g., fludarabine) are well-known risk factors for fungal diseases in general and invasive aspergillosis in particular.[6,7,37]

The impact of the underlying condition and the type of therapy are affected by other patient-related factors (Table 1). Most critical of these factors are as follows:

- The progression of the hematological malignancy is associated with poor outcome in invasive aspergillosis, while being also likely to lead to an increased risk of developing the disease.[33,38]
- More episodes of fungal infections occur in patients with acute myeloblastic leukemia during remission induction chemotherapy as compared to consolidation chemotherapy.[39]

- Higher bone marrow iron stores increase the risk of invasive aspergillosis in patients with leukemia and in hematopoietic stem transplant recipients.[32] A role of iron overload in increasing the risk of dissemination of the invasive aspergillosis has been suggested in liver transplant recipients.[40]
- Graft versus host disease in allogeneic stem cell transplant recipients requiring an increase in immunosuppressive therapy with or without corticosteroids results in a higher risk of aspergillosis.[41] Corticosteroid-refractory graft versus host disease is a major factor for developing opportunistic infections, including aspergillosis.
- A higher risk of aspergillosis is associated with allograft or renal insufficiency, transplantation or retransplantation for fulminant hepatic failure in liver transplant recipients.[42]
- Neutropenia, prior cytomegalovirus infection, and renal dysfunction are risk factors for aspergillosis in lung transplant recipients.[43]
- Renal impairment possibly also plays a role in patients with acute myeloblastic leukemia (personal unpublished data).
- Prior non-*Aspergillus* respiratory disease (such as chronic obstructive pulmonary disease, pulmonary fibrosis, residual lesion after tuberculosis or sarcoidosis, asthma, lung involvement by the malignancy, prior lung irradiation, recent respiratory virus infection. . .) was found in a high proportion of onco-hematologic

patients with invasive aspergillosis, suggesting that this condition may be a risk factor.[38,44]

- Corticosteroid therapy increases the risk of both bacterial and fungal infections in patients with chronic obstructive pulmonary diseases.[44]
- Diabetes is a well-known risk factor for mucormycosis, although this condition has not been recognized as a risk factor for aspergillosis when occurring alone. In our patient cohort with invasive aspergillosis, diabetes was present in 55 out of 303 patients (18%).[38] This rate of diabetes cases proved to be much higher than that usually encountered among patients admitted to our department of oncology and hematology, suggesting that diabetes is an additional risk factor for aspergillosis in cancer patients.

Innate immunity plays a critical role in the defense against various pathogens, including *Aspergillus* spp., in insects, while its key role has also been recognized in mammals.[45] Pathogen-associated molecular patterns (e.g., mannan, glucan, chitin, DNA, and RNA) are recognized by pattern recognition receptors (PRRs). Polymorphism in PRRs (e.g., Toll-like and C-type lectin receptors) and downstream signaling molecules, such as defensins, chemokines, cytokines, and reactive oxygen species, was reported to be associated with increased or decreased susceptibility to infections, including invasive aspergillosis.[45–47] Several studies confirmed the crucial role of innate immunity in human aspergillosis.[31,48–50]

Environmental risk factors

The patient's environment at home and during hospitalization is likely critical. Outbreaks of invasive aspergillosis were reported during construction work inside or outside of the hospital.[7] The hospitalization of highest-risk patients in units with filtered high-efficiency particulate air, with or without laminar air flow, was shown to efficiently prevent the occurrence of nosocomial invasive aspergillosis. The control of the patient's environment at home proved more difficult. Renovations, gardening, repotting house plants, activities involving exposure to dust that potentially contains fungal spores, and contact with food or spices rich in fungal spores (e.g., pepper) are all contraindicated for highly immunosuppressed patients. While restricting these activities

reduces the risk of exposure to a high amount of fungal spores, this does not prevent exposure to low counts of spores in the air, which we permanently face.

A seasonal effect was observed in a large study performed in Seattle, with patients receiving allogeneic hematopoietic stem cell transplantations during the summer being more likely to develop invasive aspergillosis.[51] Climate is known to strongly depend on geography, which may account for the large between-center differences in the incidence of invasive aspergillosis. Changes in global climatic conditions (global warming) may also exhibit an impact, which needs to be further investigated.

Personal habits, such as smoking, increases the risk by exposing the lower respiratory tract to the fungal spores contained in tobacco or cannabis.[52,53] Furthermore, the impact of the residence place, i.e., countryside versus cities or semiurban areas, has not yet been analyzed.

Risk factor or prognostic factor?

Unsurprisingly, several risk factors for invasive aspergillosis are also factors associated with a higher risk of death, the most common being the presence of corticosteroid-refractory graft versus host disease in allogeneic hematopoietic stem cell transplant recipients, severe prolonged neutropenia in leukemia patients, progression of the underlying malignancy, and corticosteroid exposure.[38,54] Impaired renal function or prior respiratory disease is very likely also a risk and prognosis factor.[38,55]

Impact on therapeutic strategies and conclusion

The early initiation of antifungal therapy is the most critical factor toward a favorable outcome. Initiating treatment as preemptive therapy prior to the mycological documentation of the disease (possible invasive aspergillosis according to the EORTC/MSG consensus criteria as opposed to probable or proven invasive aspergillosis) was reported to significantly improve the 12-week survival rate.[38] Risk stratification helps identify the patients likely to suffer from invasive aspergillosis, even if they do not present the classical risk factors, as defined in the guidelines or clinical trials. However, the multiplicity of risk factors identified in patients with invasive aspergillosis as well as the obvious synergy between these factors renders the stratification of risk factors more difficult. International, large-scale, multicenter, epidemiological studies with the inclusion of environmental and climatic conditions are necessary in order to elaborate a risk score.

Conflicts of interest

The authors declare no conflicts of interest.

References

1. Kontoyiannis, D.P., K.A. Marr, B.J. Park, *et al.* 2010. Prospective surveillance for invasive fungal infections in hematopoietic stem cell transplant recipients, 2001–2006: overview of the Transplant-Associated Infection Surveillance Network (TRANSNET) Database. *Clin. Infect. Dis.* **50:** 1091–1100.

2. Pagano, L., M. Caira, A. Candoni, *et al.* 2006. The epidemiology of fungal infections in patients with hematologic malignancies: the SEIFEM-2004 study. *Haematologica* **91:** 1068–1075.

3. Lortholary, O., S. Ascioglu, P. Moreau, *et al.* 2000. Invasive aspergillosis as an opportunistic infection in nonallografted patients with multiple myeloma: a European Organization for Research and Treatment of Cancer/ Invasive Fungal Infections Cooperative Group and the Intergroupe Francais du Myelome. *Clin. Infect. Dis* **30:** 41–46.

4. Kim, S.J., J.H. Moon, H. Kim, *et al.* 2012. Non-bacterial infections in Asian patients treated with alemtuzumab: a retrospective study of the Asian Lymphoma Study Group. *Leuk. Lymphoma* **53:** 1515–1524.

5. Pappas, P.G., B.D. Alexander, D.R. Andes, *et al.* 2010. Invasive fungal infections among organ transplant recipients: results of the Transplant-Associated Infection Surveillance Network (TRANSNET). *Clin. Infect. Dis.* **50:** 1101–1111.

6. Herbrecht, R., S. Natarajan-Ame, V. Letscher-Bru & M. Canuet. 2004. Invasive pulmonary aspergillosis. *Semin. Respir. Crit. Care Med.* **25:** 191–202.

7. Pagano, L., M. Akova, G. Dimopoulos, *et al.* 2011. Risk assessment and prognostic factors for mould-related diseases in immunocompromised patients. *J. Antimicrob. Chemother* **66(Suppl 1):** i5–i14.

8. Nosari, A., P. Oreste, R. Cairoli, *et al.* 2001. Invasive aspergillosis in haematological malignancies: clinical findings and management for intensive chemotherapy completion. *Am. J. Hematol.* **68:** 231–236.

9. Singh, N. & D.L. Paterson. 2005. Aspergillus infections in transplant recipients. *Clin. Microbiol. Rev.* **18:** 44–69.

10. Iversen, M., C.M. Burton, S. Vand, *et al.* 2007. Aspergillus infection in lung transplant patients: incidence and prognosis. *Eur. J. Clin. Microbiol. Infect. Dis.* **26:** 879–886.

11. Yan, X., M. Li, M. Jiang, *et al.* 2009. Clinical characteristics of 45 patients with invasive pulmonary aspergillosis: retrospective analysis of 1711 lung cancer cases. *Cancer* **115:** 5018–5025.

12. Garcia-Vidal, C., A. Upton, K.A. Kirby & K.A. Marr. 2008. Epidemiology of invasive mold infections in allogeneic stem cell transplant recipients: biological risk factors for infection according to time after transplantation. *Clin. Infect. Dis.* **47:** 1041–1050.

13. Zaoutis, T.E., K. Heydon, J.H. Chu, *et al.* 2006. Epidemiology, outcomes, and costs of invasive aspergillosis in immunocompromised children in the United States, 2000. *Pediatrics* **117:** e711-e716.

14. Robertson, J., O. Elidemir, E.U. Saz, *et al.* 2009. Hypogammaglobulinemia: incidence, risk factors, and outcomes following pediatric lung transplantation. *Pediatr. Transplant.* **13:** 754–759.

15. Murray, C.K., F.L. Loo, D.R. Hospenthal, *et al.* 2008. Incidence of systemic fungal infection and related mortality following severe burns. *Burns* **34:** 1108–1112.

16. Crassard, N., H. Hadden, C. Pondarre, *et al.* 2008. Invasive aspergillosis and allogeneic hematopoietic stem cell transplantation in children: a 15-year experience. *Transpl. Infect. Dis.* **10:** 177–183.

17. Silva, M.F., A.S. Ribeiro, F.J. Fiorot, *et al.* 2012. Invasive aspergillosis: a severe infection in juvenile systemic lupus erythematosus patients. *Lupus* **21:** 1011–1016.

18. Post, M.J., C. Lass-Floerl, G. Gastl & D. Nachbaur. 2007. Invasive fungal infections in allogeneic and autologous stem cell transplant recipients: a single-center study of 166 transplanted patients. *Transpl. Infect. Dis.* **9:** 189–195.

19. Kim, H.J., Y.J. Park, W.U. Kim, *et al.* 2009. Invasive fungal infections in patients with systemic lupus erythematosus: experience from affiliated hospitals of Catholic University of Korea. *Lupus* **18:** 661–666.

20. Gao, X., L. Chen, G. Hu & H. Mei. 2010. Invasive pulmonary aspergillosis in acute exacerbation of chronic obstructive pulmonary disease and the diagnostic value of combined serological tests. *Ann. Saudi. Med.* **30:** 193–197.

21. Ju, M.K., D.J. Joo, S.J. Kim, *et al.* 2009. Invasive pulmonary aspergillosis after solid organ transplantation: diagnosis and treatment based on 28 years of transplantation experience. *Transplant Proc.* **41:** 375–378.

22. Gil, L., M. Kozlowska-Skrzypczak, A. Mol, *et al.* 2009. Increased risk for invasive aspergillosis in patients with lymphoproliferative diseases after autologous hematopoietic SCT. *Bone Marrow Transplant* **43:** 121–126.

23. Wang, W., C.Y. Zhao, J.Y. Zhou, *et al.* 2011. Invasive pulmonary aspergillosis in patients with HBV-related liver failure. *Eur. J. Clin. Microbiol. Infect. Dis.* **30:** 661–667.

24. Martino, R., J.L. Pinana, R. Parody, *et al.* 2009. Lower respiratory tract respiratory virus infections increase the risk of invasive aspergillosis after a reduced-intensity allogeneic hematopoietic SCT. *Bone Marrow Transplant* **44:** 749–756.

25. Vehreschild, J.J., P.J. Brockelmann, C. Bangard, *et al.* 2012. Pandemic 2009 influenza A(H1N1) virus infection coinciding with invasive pulmonary aspergillosis in neutropenic patients. *Epidemiol. Infect.* **140:** 1848–1852.

26. Ballard, J., L. Edelman, J. Saffle, *et al.* 2008. Positive fungal cultures in burn patients: a multicenter review. *J. Burn. Care Res.* **29:** 213–221.

27. Guinea, J., M. Torres-Narbona, P. Gijon, *et al.* 2010. Pulmonary aspergillosis in patients with chronic obstructive pulmonary disease: incidence, risk factors, and outcome. *Clin. Microbiol. Infect.* **16:** 870–877.

28. Mihu, C.N., E. King, O. Yossepovitch, *et al.* 2008. Risk factors and attributable mortality of late aspergillosis after T-cell depleted hematopoietic stem cell transplantation. *Transpl. Infect. Dis* **10:** 162–167.

29. Mikulska, M., A.M. Raiola, B. Bruno, *et al.* 2009. Risk factors for invasive aspergillosis and related mortality in recipients of allogeneic SCT from alternative donors: an analysis of 306 patients. *Bone Marrow Transplant* **44:** 361–370.

30. Michallet, M., M. Sobh, S. Morisset, *et al.* 2011. Risk factors for invasive aspergillosis in acute myeloid leukemia patients prophylactically treated with posaconazole. *Med. Mycol.* **49:** 681–687.

31. Bochud, P.Y., J.W. Chien, K.A. Marr, *et al.* 2008. Toll-like receptor 4 polymorphisms and aspergillosis in stem-cell transplantation. *N. Engl. J. Med.* **359:** 1766–1777.

32. Kontoyiannis, D.P., G. Chamilos, R.E. Lewis, *et al.* 2007. Increased bone marrow iron stores is an independent risk factor for invasive aspergillosis in patients with high-risk hematologic malignancies and recipients of allogeneic hematopoietic stem cell transplantation. *Cancer* **110:** 1303–1306.

33. Cornely, O.A., J. Maertens, M. Bresnik, *et al.* 2007. Liposomal amphotericin B as initial therapy for invasive mold infection: a randomized trial comparing a high-loading dose regimen with standard dosing (AmBiLoad trial). *Clin. Infect. Dis* **44:** 1289–1297.

34. Herbrecht, R., D.W. Denning, T.F. Patterson, *et al.* 2002. Voriconazole versus amphotericin B for primary therapy of invasive aspergillosis. *N. Engl. J. Med.* **347:** 408–415.

35. Herbrecht, R., J. Maertens, L. Baila, *et al.* 2010. Caspofungin first-line therapy for invasive aspergillosis in allogeneic hematopoietic stem cell transplant patients: an European Organisation for Research and Treatment of Cancer study. *Bone Marrow Transplant* **45:** 1227–1233.

36. Viscoli, C., R. Herbrecht, H. Akan, *et al.* 2009. An EORTC Phase II study of caspofungin as first-line therapy of invasive aspergillosis in haematological patients. *J. Antimicrob. Chemother.* **64:** 1274–1281.

37. de Pauw, B., T.J. Walsh, J.P. Donnelly, *et al.* 2008. Revised definitions of invasive fungal disease from the European Organization for Research and Treatment of Cancer/Invasive Fungal Infections Cooperative Group and the National Institute of Allergy and Infectious Diseases Mycoses Study Group (EORTC/MSG) Consensus Group. *Clin. Infect. Dis.* **46:** 1813–1821.

38. Nivoix, Y., M. Velten, V. Letscher-Bru, *et al.* 2008. Factors associated with overall and attributable mortality in invasive aspergillosis. *Clin. Infect. Dis.* **47:** 1176–1184.

39. Cordonnier, C., C. Pautas, S. Maury, *et al.* 2009. Empirical versus preemptive antifungal therapy for high-risk, febrile, neutropenic patients: a randomized, controlled trial. *Clin. Infect. Dis.* **48:** 1042–1051.

40. Singh, N. & H.Y. Sun. 2008. Iron overload and unique susceptibility of liver transplant recipients to disseminated disease due to opportunistic pathogens. *Liver Transplant.* **14:** 1249–1255.

41. Marr, K.A., R.A. Carter, M. Boeckh, *et al.* 2002. Invasive aspergillosis in allogeneic stem cell transplant recipients: changes in epidemiology and risk factors. *Blood* **100:** 4358–4366.

42. Hellinger, W.C., H. Bonatti, J.D. Yao, *et al.* 2005. Risk stratification and targeted antifungal prophylaxis for prevention of aspergillosis and other invasive mold infections after liver transplantation. *Liver Transplant.* **11:** 656–662.

43. Gordon, S.M. & R.K. Avery. 2001. Aspergillosis in lung transplantation: incidence, risk factors, and prophylactic strategies. *Transpl. Infect. Dis.* **3:** 161–167.

44. Cornillet, A., C. Camus, S. Nimubona, *et al.* 2006. Comparison of epidemiological, clinical, and biological features of invasive aspergillosis in neutropenic and nonneutropenic patients: a 6-year survey. *Clin. Infect. Dis.* **43:** 577–584.

45. Romani, L. 2011. Immunity to fungal infections. *Nat. Rev. Immunol.* **11:** 275–288.

46. Lamoth, F., I. Rubino & P.Y. Bochud. 2011. Immunogenetics of invasive aspergillosis. *Med. Mycol.* **49**(**Suppl 1**): S125–S136.

47. Bochud, P.Y., M. Bochud, A. Telenti & T. Calandra. 2007. Innate immunogenetics: a tool for exploring new frontiers of host defence. *Lancet Infect. Dis.* **7:** 531–542.

48. Lambourne, J., D. Agranoff, R. Herbrecht, *et al.* 2009. Association of mannose-binding lectin deficiency with acute invasive aspergillosis in immunocompromised patients. *Clin. Infect. Dis.* **49:** 1486–1491.

49. Sainz, J., C.B. Lupianez, J. Segura-Catena, *et al.* 2012. Dectin-1 and DC-SIGN polymorphisms associated with invasive pulmonary Aspergillosis infection. *PLoS One* **7:** e32273.

50. Zaas, A.K., G. Liao, J.W. Chien, *et al.* 2008. Plasminogen alleles influence susceptibility to invasive aspergillosis. *PLoS Genet.* **4:** e1000101.

51. Panackal, A.A., H. Li, D.P. Kontoyiannis, *et al.* 2010. Geoclimatic influences on invasive aspergillosis after hematopoietic stem cell transplantation. *Clin. Infect. Dis.* **50:** 1588–1597.

52. Szyper-Kravitz, M., R. Lang, Y. Manor & M. Lahav. 2001. Early invasive pulmonary aspergillosis in a leukemia patient linked to aspergillus contaminated marijuana smoking. *Leuk. Lymphoma* **42:** 1433–1437.

53. Verweij, P.E., J.J. Kerremans, A. Voss & J.F. Meis. 2000. Fungal contamination of tobacco and marijuana. *JAMA* **284:** 2875.

54. Ribaud, P., C. Chastang, J.P. Latge, *et al.* 1999. Survival and prognostic factors of invasive aspergillosis after allogeneic bone marrow transplantation. *Clin. Infect. Dis.* **28:** 322–330.

55. Upton, A., K.A. Kirby, P. Carpenter, *et al.* 2007. Invasive aspergillosis following hematopoietic cell transplantation: outcomes and prognostic factors associated with mortality. *Clin. Infect. Dis.* **44:** 531–540.

Ann. N.Y. Acad. Sci. ISSN 0077-8923

Invasive aspergillosis in the intensive care unit

George Dimopoulos,[1] Frantezeska Frantzeskaki,[1] Garyfallia Poulakou,[2] and Apostolos Armaganidis[1]

[1]Department of Critical Care, [2]Fourth Department of Internal Medicine, University Hospital ATTIKON, Medical School, University of Athens, Greece

Address for correspondence: George Dimopoulos, M.D., Ph.D., F.C.C.P., Department of Critical Care, University Hospital ATTIKON, Medical School, University of Athens, 1 Rimini str., 14569-Haidari, Athens, Greece. gdimop@med.uoa.gr

Invasive aspergillosis is a devastating infection affecting severely immunocompromised patients, most frequently with hematologic malignancies. In recent years, a surge in the incidence of invasive aspergillosis has been reported in critically ill patients without the classical risk factors. The mortality of the disease is equally high in the group of intensive care unit (ICU) patients, while the clinical signs and symptoms are nonspecific and the diagnosis remains a challenge. New noninvasive diagnostic methods in combination with better tolerated antifungal drugs aim for early diagnosis and improved prognosis of invasive aspergillosis. In this review, we discuss the epidemiology, the diagnostic strategy and algorithms, and the therapeutic choices of this severe infection in critically ill patients.

Keywords: aspergillosis; diagnosis; treatment

Introduction

Systemic infections are a common cause of death in critically ill patients, and pulmonary aspergillosis (PA) ranks among the most lethal and difficult diagnosed infections in this population. *Aspergillus* spp. can cause a broad spectrum of lung infections, manifesting as acute invasive pulmonary aspergillosis (IA), local necrosis with cavitation, mycetoma, and tracheobronchitis of allergic bronchopulmonary aspergillosis (ABPA).[1] Among these entities, IA is one of the most devastating forms of *Aspergillus* infection, with increasing frequency over the last decades. Invasive pulmonary aspergillosis (IPA), characterized by invasion and necrosis of lung parenchyma by *Aspergillus*, affects not only immunocompromised but also critically ill patients who do not have classical risk factors (for example, hematological malignancies or allogeneic hematopoietic stem cell transplantion (HSCT).[2] New risk factors that predispose to IPA in critically ill patients include chronic obstructive pulmonary disease (COPD), chronic use of systemic and inhaled corticosteroid, cirrhosis, solid-organ transplantation, and severe sepsi.[3,4] The mortality rate is high in the classic risk groups, exceeding 50% in neutropenic patients and 90% in HSCT recipients.[5,6]

Interestingly, in less immunocompromised ICU patients, the mortality rate is equally high, reaching 90%.[7]

The clinical presentation of IA is nonspecific, and the diagnostic criteria are poorly defined in critically ill populations, rendering the diagnosis of the disease challenging. A high index of clinical suspicion is necessary; for example, the isolation of *Aspergillus* spp. in lower respiratory tract samples should not always be considered to indicate colonization.[8] New diagnostic tests have been introduced in daily practice in order to maximize the chance for therapy. This review covers the epidemiology, risk factors, diagnostic procedures and algorithms, and therapeutic choices of invasive aspergillosis in ICU patients.

Epidemiology and risk factors

Aspergillus fumigatus, A. flavus, A. niger, A. terreus, and *A. nidulans* are the most common isolated *Aspergillus* spp. in humans.[9,10] The first description of IA as an opportunistic infection was published in 1953.[4] An increasing incidence has been observed during the past two decades and attributed to the use of chemotherapeutic and immunosuppressive drugs. The most important risk factors for

doi: 10.1111/j.1749-6632.2012.06805.x

Table 1. Reported incidence of aspergillosis in the ICU setting

Author	Year	Incidence (%)
Roosen	2000	15
Valles	2002	19
Meersseman	2003	5.8
Dimopoulos	2004	3.7
Garnacho-Montero	2005	1.1
Kumar	2006	0.7
Vandewoude	2006	0.3

IPA are the various types of immunodeficiency, including HSCT, prolonged and severe neutropenia (more than three weeks with neutrophils $< 500/\mu$L), hematologic malignancies, corticosteroid therapy, and HIV infection.[11] Neutropenia is the most significant risk factor, since neutrophils and alveolar macrophages are the most important host defenses against *Aspergillus*.[12] Several cases of IPA following HSCT have recently been described. Patients after allogeneic HSCT are in higher risk for IPA comparing to autologous HSCT (frequencies 2.3–15% and 0.5–4%, respectively).[2]

However, there are several reports of IA in immunocompetent patients who are critically ill without neutropenia, hematological malignancy, or HSCT. The incidence of IA in a typical ICU varies from 0.33% to 6.9% (Table 1). Interestingly, histopathological or microbiological evidence of IA has been observed in 6.9% of ICU patients in a retrospective autopsy-controlled study, although most (70%) did not have a diagnosis of hematological malignancy, a classical risk factor of IA and the reported mortality was 90%.[7] According to other studies, IA is one of the most frequently undiagnosed infections in the critically ill.[13] Mortality rates in ICU patients with proven or probable IPA (Fig. 1) vary between 59% and 95%, depending on the relevant study (Fig. 1).[14]

The risk factors for invasive aspergillosis among patients admitted to the ICU can be classified in three categories: *high risk* includes patients with neutropenia, hematological malignancy, or allogeneic HSCT; *intermediate risk* includes patients undergoing prolonged treatment with corticosteroids, those who have undergone autologous HSCT, and patients with COPD, liver chirrosis, solid-organ cancer, HIV infection, lung transplantation, or sys-temic immunosuppresive therapy; and *low risk* consists of patients with severe burns, solid-organ transplantation, steroid treatment (more than seven days), a prolonged ICU stay (more than 21 days), multiple organ dysfunction syndrome, malnutrition, postcardiac surgery, and near drowning.[1] In addition to the classical high-risk groups, three non-classical risk groups have also been recognized for IP: patients with COPD, liver cirrhosis, or disturbed immunoregulation due to critical illness. It is well known that the systematic inflammatory response syndrome (SIRS) leads to anti-inflammatory response, characterized by a decrease of innate and adaptive immunity, a decrease of all cellular immune functions, and a decrease of the number of myeloid dendritic cells.[15]

The incidence of IPA in COPD patients is 1.3%, according to a review of 50 studies.[16] According to Bulpa *et al.*, IPA in this group of patients is defined as proven, probable, possible, or colonization based on histopathological, clinical, and radiological criteria (Table-slide 13 in Bulpa). This classification is based on the criteria of the European Organization for Research and Treatment of Cancer/Invasive Fungal Infections Cooperative Group and the National Institutes of Allergy and Infectious Disease Mycoses Study Group (EORTC/MSG) for the diagnosis of invasive fungal infections in immunocompromised patients.[17] The emergence of IPA in this group of patients has been mainly attributed to the prolonged administration of high doses of corticosteroids.[18] Nevertheless, IPA has been described in COPD patients who have never received steroids, while some viral infections may play a causal role.[19] Guinea *et al.* retrospectively classified, according to Bulpa

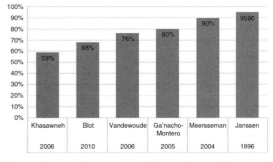

Figure 1. Mortality rates in ICU patients with proven/probable IA.

criteria, a large series of COPD patients with isolation of *Aspergillus* from respiratory samples.[20] They reported probable IPA in at least 22.1% of these patients, and logistic regression showed the following variables as independent predictors of IPA: admission to the ICU, chronic heart failure, recent antibiotic treatment, and administration of high doses of steroids.

Patients with severe liver disease are also at high risk for IA, with mortality rates exceeding 70%.[21] These patients, though apparently immunocompetent, are prone to bacterial or fungal infections as both humoral and cell-mediated immunity are depressed by liver dysfunction, opsonization is decreased due to complement deficiencies, neutrophil migration and phagocytosis are defected, and the oxidative burst activity of neutrophils is impaired.[13,22–24] The immune system of these patients is additionally weakened because of probable corticosteroid therapy, transfusions of allogeneic blood products, hemodialysis, and sepsis.[13] Critically ill patients are prone to IPA in the absence of hematologic malignancy, neutropenia, COPD, or liver disease, as many critical illnesses have concomitant disturbances in immunoregulation characterized by macrophage deactivation and an altered cellular immune response termed *immunoparalysis*.[25] Moreover, other predisposing factors such as hyperglycemia, corticosteroid use, renal failure and continuous renal replacement therapy, and broad spectrum antibiotic treatment may lead to invasive fungal disease.[26,27] Concerning the pathogenesis of IPA, *Aspergillus* has a sporulating capacity ranging from 1 to 100 spores/mm^3 in the normal air.[28,29] The conidia released from the spores are small enough (2–3 μm of diameter) to reach the alveoli by inhalation, thus the respiratory tract is the main source of entry. The impaired immunologic response of the host may lead to *Aspergillus* colonization of the respiratory system and subsequently to invasive aspergillosis.

Diagnosis

The diagnosis of IPA is challenging, especially in the ICU setting. The diagnostic methods can be classified as laboratory and high technology. Laboratory methods are the microscopic examination of various samples, culture and identification of the fungus, histopathological examination of tissues, antibody detection, antigen detection, and PCR.[30]

High technology methods include the detection of antibodies and metabolites of fungus, detection of fungal cell wall components, and fungal PCR. The gold standard of diagnosis remains the histopathological examination of invasive lung tissue sampling (thoracoscopic or open lung biopsy).[31] The presence of nonpigmented septate hyphae with repeated dichotomous branching and a culture positive for *Aspergillus* from the same site is diagnostic. However, a proven diagnosis in ICU patients is very difficult, since coagulopathy and hemodynamic instability render invasive procedures rather difficult. Therefore, several alternative diagnostic methods and algorithms have been developed.

In recent years there is an increasing interest in the detection of *Aspergillus* spp. antigens or DNA in body fluids (e.g., serum, bronchoalveolar lavage (BAL), and cerebrospinal fluid (CSF)). Galactomannan and (1,3)-β-D-glucan are fungal cellular wall components frequently used for the IA diagnosis. Galactomannan (GM), a polysaccharide released in body fluids during *Aspergillus* growth, can be detected in serum and BAL by a double-sandwich ELISA technique several days before clinical and radiological signs of the disease become apparent.[2] Maertens *et al.* compared the accuracy of the galactomannan assay at multiple cutoffs and concluded that consecutive positive results at a threshold of 0.5 ng/mL had best sensitivity and specificity, but that a single result of >0.8 was also basis for therapy, with sensitivity of 96% and specificity of 97%.[32] According to other reports, GM detection had a sensitivity of 71% and specificity of 89% for IPA in patients with a hematological malignancy or HSCT, but the method is less useful in nonneutropenic patients.[33]

Meersseman *et al.* reported the diagnostic significance of GM measurement in BAL of critically ill patients;[7] in this study 110 patients were included, 36 with hematologic malignancy and 74 with other immunocompromising factors. The sensitivity of GM detection was 88% with a cut off value of 0.5, compared to 42% for the respective measurement in serum, while the specificity of GM in both BAL and serum was 86%. The low sensitivity of serum GM levels for the detection of IP, especially in nonneutropenic patients, is attributed to the clearing of GM by neutrophils from the circulation.[34] Additionally the serum of BAL GM levels may be falsely elevated in patients receiving either dietary sources of GM (cereals, pasta) or several antibiotics (for example, piperacillin-tazobactam).[34,35]

Another fungal cell wall constituent, highly sensitive and specific for invasive mycosis, is (1,3)-β-D-glucan. The sensitivity of this assay for IPA diagnosis varies from 55% to 100% and the specificity from 52% to 100%, with false positive results observed in patients with bacterial infections, liver disease, hemodialysis, abdominal surgery, and antibiotics.[9] The same antigen might be present in other fungal infections (e.g., candidiasis). The combination of galactomannan and (1,3)-β-D-glucan might improve specificity and positive predictive value of each separate test.[36] Fungal DNA can be detected in blood, serum, tissue biopsies, BAL, and CSF by application of molecular techniques, in particular PCR. The sensitivity and specificity of the method for BAL are 67–100% and 55–95%, respectively, while the respective values for serum are 100% and 65–92%.[37,38] The main problem with PCR is that specimens can be contaminated; in addition, PCR cannot discriminate colonization from true infection. Although PCR facilitates earlier the possible diagnosis of IPA, it is not a routine clinical assay; and no study has proven that it has a significant impact on patient management or outcome.

In the early stages of IPA, chest radiograph is usually negative or shows nonspecific changes. Chest high-resolution computer tomography (HRCT) is useful for the early diagnosis of IPA. Early lesions include multiple nodules and the characteristic halo sign, which appears mainly in neutropenic patients as a zone of low attenuation due to coagulation necrosis with decreased blood flow in the center that is surrounded by a translucent ground-glass halo.[39] Later, in the course of the disease after the recovery of neutrophil function, the air crescent sign, or central cavitation, is apparent and characterized by coagulation necrosis of an original nodule.[40] None of these signs is pathognomonic for IPA, and according to Vanderwoude *et al.* the most common radiological findings in ICU patients with IPA are nonspecific infiltrates and consolidation.[8] The halo sign can also be found in infections caused by *Nocardia* spp, Mucorales, *Trichosporon* spp, and *Fusarium* spp, and in patients with metastases, eosinophilic pneumonia, or bronchoalveolar carcinoma.[41,42]

According to the EORTC/MSG criteria, the definition of proven aspergillosis includes the identification of *Aspergillus* by histopathological examination or culture. Probable disease requires at least one immunocompromising host factor, one clinical criterion, and mycological evidence (culture or antigen). In the absence of mycological evidence, IA is defined as possible.[43] These definitions, although expanded to include nonimmunocompromised hosts, have poor diagnostic accuracy in ICU patients.[8] The isolation of *Aspergillus* from respiratory samples in critically ill patients is associated with IPA in 10–80% of them, depending on the study.[44,45] According to Perfect *et al.*, 12% of patients with *Aspergillus* species isolated from respiratory samples developed invasive aspergillosis. Furthermore, the isolation of *Aspergillus* from respiratory samples carried a poor prognosis, as only 38% of the patients were alive three months after diagnosis.[46] Dimopoulos *et al.* reported that 2.7% of 222 ICU patients had evidence of disseminated aspergillosis at autopsy; *A. fumigatus* was isolated from respiratory samples of all these patients premortem and considered to be colonization.[26] Interestingly, Khasawneh *et al.* performed a retrospective review of medical reports to determine the clinical significance of isolation of *Aspergillus* from respiratory samples of critically ill patients and found that 28% of them developed probable or definite IA, and 78% of them were *Aspergillus* colonization cases.[14] However, Khasawneh *et al.* concluded that the isolation of the fungus is a poor prognostic marker irrespective of invasive disease or colonization.

Given all these diagnostic problems, Vandewoude *et al.* developed an adapted clinical algorithm based on the EORTC/MSG criteria in order to discriminate *Aspergillus* colonization from invasive disease in critically ill patients, with isolation of the fungus from endotracheal aspirates: 48% of the patients were classified as IPA and 52% as colonization.[8] The available histopathology data from 26 patients confirmed the diagnosis based upon the clinical algorithm. Blot *et al.* externally validated this clinical diagnostic algorithm in order to discriminate colonization from IPA in mechanically ventilated critically ill patients, with *Aspergillus* isolated from endotracheal aspirates (Table 2).[47] It is worth noting that the evaluated algorithm accepted any radiological abnormality, since radiological lesions are nonspecific in ICU patients. Additionally, the algorithm contained a culture and microscopy positive BAL as an alternative criterion for host factors, since IPA

Table 2. Criteria proposed for the diagnosis of probable IPA in the ICU setting[48]

---

1. *Aspergillus (+)* in lower respiratory tract specimen culture (entry criterion)
2. Compatible signs and symptoms (one of the following)
  • Fever refractory to at least three days of appropriate antibiotic therapy
  • Recrudescent fever after a period of defervescence of at least 48 h while still on antibiotics
    and without other apparent cause
  • Pleuritic chest pain
  • Pleuritic rub
  • Dyspnea
  • Hemoptysis
  • Worsening respiratory insufficiency in spite of appropriate antibiotic therapy and ventilatory
    support
3. Abnormal medical imaging by portable chest X-ray or CT scan of the lungs
4. Either
  4a. Host risk factors (one of the following conditions)
  • Neutropenia (absolute neutrophil count less then 500/mm^3) preceding or at the time of ICU
    admission
  • Underlying hematological or oncological malignancy treated with cytotoxic agents
  • Glucocorticoid treatment (prednisone or equivalent, >20 mg/day)
  • Congenital or acquired immunodeficiency
  OR
  4b. Semiquantitative *Aspergillus*-positive
  • Culture of BAL fluid (+ or ++) without bacterial growth together with a positive cytological
    smear showing branching hyphae
Probably invasive pulmonary aspergillosis: 1 + 2 + 3 + either 4a or 4b
When ≥1 criterion is not met, the case is classified as *Aspergillus* colonization

---

may develop in patients without apparent risk factors. IPA was defined as probable, if all the criteria were fulfilled. The case was defined as colonization if one or more criteria were not met. The judgment, based upon the algorithm in each case, was compared to the results of histopathology examination on lung biopsy or autopsy. The algorithm demonstrated 61% specificity and 91% sensitivity for the discrimination of IPA from colonization in mechanically ventilated patients.

Treatment

Treatment of IPA remains difficult, although new antifungal agents have been introduced in clinical practice. Amphotericin B has been the standard therapy of IPA for decades. However, many studies have shown reduced efficacy of the drug, partially due to the development of resistance among *Aspergillus* subspecies and the serious adverse effects associated with its use, in particular nephrotoxicity, electrolyte disturbances, and hypersensitivity reac-

tions.[48,49] Newer lipid-based preparations of amphotericin with reduced nephrotoxicity have been introduced in the treatment of fungal infections. Nevertheless, these agents are more expensive and higher doses may be needed for the same antifungal therapeutic result.[50]

The treatment of choice for IA is voriconazole, a broad-spectrum triazole.[51,52] In a large, prospective, multicenter trial Herbrecht *et al.* compared voriconazole to amphotericin B for the treatment of IPA in patients with hematologic diseases.[53] At week 12 a significantly higher rate of favorable response was observed in patients receiving voriconazole (53%) compared to patients receiving amphotericin B (32%), as well as a higher 12-week survival (71% vs. 58%). Fewer drug-related side effects were reported with voriconazole. The drug is available in intravenous (i.v.) and oral formulations, with a recommended dose of 6 mg/kg twice daily i.v. on day 1 and then 4 mg/kg twice daily. The clearance of the drug is hepatic by the cytochrome P450, so the possibility of drug–drug interaction is considerable: voriconazole interacts with many drugs, included, but not limited to, cyclosporine, warfarin, carbamazepine, and statins, and it reduces the clearance of midazolame.[51] Additionally, fatal interaction with highly active antiretroviral therapy (HAART) has been reported in patients with HIV infection.[54] However, voriconazole is generally well tolerated. The most commonly described adverse effect is visual disturbances, such as blurred vision, photophobia, and altered color perception. Less frequent toxicities include liver function test abnormalities and dermatologic reactions.[2] The intravenous form of the drug includes the solvent vehicle sulfobutylether beta cyclodextrin sodium,[55] which accumulates in cases of moderate and severe renal impairment with possible toxic effects on kidney and liver. Therefore, it is recommended that only the oral form of voriconazole be used in cases of renal dysfunction. However, in critically ill patients the oral administration of the drug may lead to insufficient intestinal absorption.[56] Monitoring of therapeutic blood levels of voriconazole may reduce the unnecessary discontinuation of the drug due to adverse effects and improve the therapeutic response in invasive aspergillosis.[57]

Echinocandins, such as caspofungin, micafungin, and anidulafungin, could be alternative drugs for the treatment of IPA in refractory cases. Caspofungin has proven to be effective in the treatment of IPA in patients who cannot tolerate the first-line agents or in cases of IPA refractory to standard treatment.[58] Caspofungin as a first-line treatment for IPA, in contrast, has not been compared to voriconazole. Echinocandins inhibit the $(1,3)$-β-D-glucan component of the fungal cell wall, while azoles target the cell membrane. Therefore, combination therapy might be considered in refractory cases of IPA.[59] Potential advantages could be the more rapid fungicidal activity, the reduction of the doses of individual agents, and the prevention of emergence of resistance.[60] There are few clinical data concerning combination antifungal therapy and most of the studies have important limitations. Nevertheless, studies do not support the combination of voriconazole with amphotericin B.[61] Additionally the inhibition of ergosterol synthesis by azoles deprives amphotericin B of its target, leading to an antagonistic effect of the combination.[62] The combination of caspofungin and liposomal amphotericin B showed a response rate of 42%, while a survival advantage of voriconazole plus caspofungin, compared to voriconazole alone, was found in a retrospective analysis of IPA treatment.[63,64] Particularly in the critically ill, the combination of echinocandin with a lipid formulation of amphotericin B or azole might be useful, but randomized controlled studies are needed for further evaluation of the treatment.[64] Nevertheless, combination treatment might be considered in refractory cases of IPA or in breakthrough infections.[65]

Failure of antifungal treatment could be due to intrinsic or secondary antifungal resistance. Intrinsic resistance to various antifungals agents has been reported particularly for non-*fumingatus Aspergillus* species. *A. terreus* shows primary resistance to amphotericin B, while there are also recent reports of resistance in several species of the *Fumigati* group.[66,67] Acquired resistance in *Aspergillus* species to azoles is rarely reported, occuring mainly in *A. fumigatus*.[68] There are few reports of resistance to echinocandin, in particular of echinocandin-exposed patients.[69] The clinical and radiological response of each patient to antifungal treatment will determine the duration of therapy,[70] which can be very prolonged, ranging from several months to more than one year. The clinical and radiological resolution, including the absence of fever and the decrease of thoracic aspergillary lesions on CT,[71] are prerequisites for the cessation of antifungal therapy. Regarding laboratory indices, Boutboul *et al.*, studied the kinetics of serum *Aspergillus* GM in HSCT patients and concluded that an increase of GM level 1.0 over baseline during the first week of therapy was predictive of treatment failure (with sensitivity of 44% and specificity of 87%).[72]

Environmental sources and prevention of IPA

Environmental factors combined with host factors may play a significant role in the development of IPA. The most frequent source of infection is inhalation of *Aspergillus* spores that are found in soil, decomposing plant matter, household dust, and some building materials.[73] Construction or renovation work near hospitals has been associated with nosocomial outbreaks of IPA[74,75] due to *Aspergillus* spores contaminating the hospital air. According to other reports, *A. fumigatus* can contaminate hospital water, with potential implications on patients' health.[76] However, the likelihood of hospital water as an environmental source for IPA is not clear; large trials with molecular techniques are required in order to determine its importance.

Environmental control measures are necessary in order to protect high-risk patients from exposure to *Aspergillus* spores. During hospital building work the following barrier protection methods have been used:[77] portable high-efficiency particulate air (HEPA) filter air purifier units installed in rooms housing immunocompromised individuals; a copper-8 quinolinolate formulation applied to surfaces in the rooms and above the false ceilings; windows sealed; regular cleaning of the rooms; and patients moved to other areas of the hospital until the implementation of containment measures. Other control strategies include disinfection of water supplies and methods to decrease patient exposure. However, the cost-effectiveness of these measures has to be examined, and large epidemiological studies are required in order to lead to effective prevention strategies against this lethal disease.

Conclusions

IA is an infection with poor prognosis, probably fatal in the case of delayed diagnosis. ICU patients lacking the classical risk factors can develop IA, particularly if COPD or severe hepatic failure is present. The diagnosis of the disease is rather difficult because biopsies are not easily performed in such patients. Therefore, the finding of *Aspergillus* in respiratory tract samples should raise the suspicion of IA. Critical care providers need algorithms in order to make therapeutic decisions, given the availability of new and effective drugs. Randomized controlled strategies are necessary for the validation of these algorithms and the subsequent development of guidelines and recommendations.

Conflicts of interest

The authors declare no conflicts of interest.

References

1. Dutkiewicz, R. & A. Hage. 2010. *Aspergillus* infections in the critically ill. *Proc. Am. Thorac. Soc.* **7:** 204–209.
2. Kousha, M., R. Tadi & O. Soubani. 2011. Pulmonary aspergillosis: a clinical review. *Eur. Respir. Rev.* **20:** 156–174.
3. Leav, A., B. Fanburg & S. Hadley. 2000. Invasive pulmonary aspergillosis associated with high-dose inhaled fluticasone. *N. Engl. J. Med* **343:** 586.
4. Rankin, E. 1953. Disseminated aspergillosis and moniliasis associated with agranulocytosis and antibiotic therapy. *Br. Med. J.* **1:** 918–919.
5. Yeghen, T., C. Kibbler, G. Prentice, *et al.* 2000. Management of invasive pulmonary aspergillosis in hematology patients: a review of 87 consecutive cases at a single institution. *Clin. Infect. Dis.* **31:** 859–868.
6. Fukuda, T., M. Boeckh, A. Carter, *et al.* 2003. Risks and outcomes of invasive fungal infections in recipients of allogeneic hematopoietic stem cell transplants after nonmyeloablative conditioning. *Blood* **102:** 827–833.
7. Meersseman, W., J. Vandecasteele, A. Wilmer, *et al.* 2004. Invasive aspergillosis in critically ill patients without malignancy. *Am. J. Respir. Crit. Care. Med.* **170:** 621–625.
8. Vandewoude, H., S.I. Blot, P. Depuydt, *et al.* 2006. Clinical relevance of *Aspergillus* isolation from respiratory tract samples in critically ill patients. *Crit. Care* **10:** R31.
9. Dimopoulos, G. & I. Karampela. 2009. Pulmonary Aspergillosis. Different diseases for the same pathogen. *Clin. Pulm. Med.* **16:** 1–6.
10. Denning, D. *Aspergillus* species. 2000. In *Principles and Practice of Infectious Disease*. G. Mandell, A. Douglas & J.E. Bennett, Eds.: 2675–2685. Churchill Livingstone. Philadelphia, PA.
11. Bartlett, J.C. 2000. Aspergillosis update. *Medicine* **79:** 281–282.
12. Garnacho-Montero, J. & R. Amaya-Villar. 2006. A validated clinical approach for the management of aspergillosis in critically ill patients: ready, steady, go! *Crit. Care* **10:**132.
13. Meersseman, W., K. Lagrou, J. Maertens & E. Van Wijngaerden. 2007. Invasive aspergillosis in the intensive care unit. *Clin. Infect. Dis.* **45:** 205–216.
14. Khasawneh, F., T. Mohamad, K. Moughrabieh, *et al.* 2006. Isolation of *Aspergillus* in critically ill patients: a potential marker of poor outcome. *J. Crit. Care* **21:** 322–327.
15. Stevens, A. & L. Melikian. 2011. Aspergillosis in the 'nonimmunocompromised' host. *Immunol. Invest.* **40:** 751–766.
16. Lin, S.J., J. Schranz & S.M. Teutsch. 2001. Aspergillosis casefatality rate: systematic review of the literature. *Clin. Infect. Dis.* **32:** 358–366.
17. Ascioglu, S., J.H. Rex, B. de Pauw, *et al.* 2002. Defining opportunistic invasive fungal infections in immunocompromised patients with cancer and hematopoietic stem cell transplants: an international consensus. *Clin. Infect. Dis.* **34:** 7–14.
18. Ader, F., S. Nseir, R. Le Berre, *et al.* 2005. Invasive pulmonary aspergillosis in chronic obstructive pulmonary disease: an emerging fungal pathogen. *Clin. Microbiol. Infect.* **11:** 427–429.
19. Ali, Za, A. Ali, E. Tempest & J. Wiselka. 2003. Invasive pulmonary aspergillosis complicating COPD disease in an immunocompetent patient. *J. Postgrad. Med.* **49:** 78–80.
20. Guinea, J., M. Torres-Narbona, P. Gijón, *et al.* 2010. Pulmonary aspergillosis in patients with COPD disease: incidence, risk factors, and outcome. *Clin. Microbiol. Infect.* **16:** 870–877.
21. Falcone, M., P. Massetti, A. Russo, *et al.* 2011. Invasive aspergillosis in patients with liver disease. *Med. Mycol.* **49:** 406–413.
22. Cheruvattath, R. & V. Balan. 2007. Infections in patients with end-stage liver disease. *J. Clin. Gastroent.* **41:** 403–411.

23. Stevens, A. & G. Melikian. 2011. Aspergillosis in the 'non-immunocompromised' host. *Immunol. Invest.* **40:** 751–766.

24. Fiuza, C., M. Salcedo, G. Clemente & M. Tellado. 2000. *In vivo* neutrophil dysfunction in cirrhotic patients with advanced liver disease. *J. Infect. Dis.* **182:** 526–533.

25. Hartemink, K.J., M.A. Paul, J.J. Spijkstra, *et al.* 2003. Immunoparalysis as a cause for invasive aspergillosis? *Intensive Care Med.* **29:** 2068–2071.

26. Dimopoulos, G., M. Piagnerelli, J. Berré, *et al.* 2003. Disseminated aspergillosis in ICU patients: an autopsy study. *J. Chemother.* **15:** 71–75.

27. Ng, T.T., G.D. Robson & D.W. Denning. 1994. Hydrocortisone-enhanced growth of *Aspergillus* spp.: implications for pathogenesis. *Microbiology* **140**(Pt 9): 2475–2479.

28. Streifel, J., J.L. Lauer, D. Vesley, *et al.* 1983. *Aspergillus* fumigatus and other thermotolerant fungi generated by hospital buiding demolition. *Applied Environ. Microbiol.* **46:** 375–378.

29. Latge, P. 2001. The pathobiology of *Aspergillus* fumigatus. *Trends Microbiol.* **9:** 382–389.

30. van Burik, A., R. Colven & H. Spach. 1998. Cutaneous aspergillosis. *J. Clin. Microbiol.* **36:** 3115–3121.

31. Ruhnke, M., A. Bohme, D. Buchheidt, *et al.* 2003. Diagnosis of invasive fungal infections in hematology and oncology—guidelines of the infectious diseases working party (AG-IHO) of the German society of Hematology and Oncology (DGHO). *Ann. Hematol.* **82**(Suppl 2): S141–S148.

32. Maertens, J., K. Theunissen, E. Verbeken, *et al.* 2004. Prospective clinical evaluation of lower cut-offs for galactomannan detection in adult neutropenic cancer patients and haematological stem cell transplant recipients. *Br. J. Haematol.* **126:** 852–860.

33. Pfeiffer, C.D., J.P. Fine & N. Safdar. 2006. Diagnosis of invasive aspergillosis using a galactomannan assay: a meta-analysis. *Clin. Infect. Dis.* **42:** 1417–1427.

34. Herbrecht, R., V. Letscher-Bru, C. Oprea, *et al.* 2002. *Aspergillus* galactomannan detection in the diagnosis of invasive aspergillosis in cancer patients. *J. Clin. Oncol.* **20:** 1898–1906.

35. Singh, N., A. Obman, S. Husain, *et al.* 2004. Reactivity of platelia *Aspergillus* galactomannan antigen with piperacillin-tazobactam: clinical implications based on achievable concentrations in serum. *Antimicrob. Agents Chemother.* **48:** 1989–1992.

36. Meersseman, W., K. Lagrou, J. Maertens, *et al.* 2008. Galactomannan in bronchoalveolar lavage fluid: a tool for diagnosing aspergillosis in intensive care unit patients. *Am. J. Respir. Crit. Care Med.* **177:** 27–34.

37. Hizel, K., N. Kokturk, A. Kalkanci, *et al.* 2004. Polymerase chain reaction in the diagnosis of invasive aspergillosis. *Mycoses* **47:** 338–342.

38. Buchheidt, D., C. Baust, H. Skladny, *et al.* 2001. Detection of *Aspergillus* species in blood and bronchoalveolar lavage samples from immunocompromised patients by means of 2-step polymerase chain reaction: clinical results. *Clin. Infect. Dis.* **33:** 428–435.

39. Kuhlman, J.E., E.K. Fishman & S.S. Siegelman. 1985. Invasive pulmonary aspergillosis in acute leukemia: characteristic findings on CT, the CT halo sign, and the role of CT in early diagnosis. *Radiology* **157:** 611–614.

40. Curtis, M., J. Smith & E. Ravin. 1979. Air crescent sign invasive aspergillosis. *Radiology* **133:** 17–21.

41. Gefter, B., M. Albelda, H. Talbot, *et al.* 1985. Invasive pulmonary aspergillosis and acute leukemia. Limitations in the diagnostic utility of air crescent sign. *Radiology* **157:** 605–610.

42. Gaeta, M., A. Blandino, E. Scribano, *et al.* 1999. Computed tomography halo sign in pulmonary nodules: frequency and diagnostic value. *J. Thorac. Imaging* **14:** 109–113.

43. De Pauw, B., J. Walsh, P. Donnelly, *et al.* 2008. Revised definitions of invasive fungal disease from the European Organization for Research and Treatment of Cancer/Invasive Fungal Infections Cooperative Group and the National Institute of Allergy and Infectious Diseases Mycoses Study Group (EORTC/MSG) Consensus Group. *Clin. Infect. Dis.* **46:** 1813–1821.

44. Soubani, O. & H. Chandrasekar. 2002. The clinical spectrum of pulmonary aspergillosis. *Chest* **121:** 1988–1999.

45. Fukuda, T., M. Boeckh, R.A. Carter, *et al.* 2003. Risks and outcomes of invasive fungal infections in recipients of allogeneic hematopoietic stem cell transplants after non-myeloablative conditioning. *Blood* **102:** 827–833.

46. Perfect, R., M. Cox, J. Lee, *et al.* Mycoses Study Group. 2001. The impact of culture isolation of *Aspergillus* species: a hospital-based survey of aspergillosis. *Clin. Infect. Dis.* **33:** 1824–1833.

47. Blot, S., F. Taccone, M. Van den Abeele, *et al.* 2012. A clinical algorithm to diagnose invasive pulmonary agpergillosis in critically ill. *Am. J. Respir. Crit. Care Med.* **186:** 56–64.

48. Ostrosky-Zeichner, L., A. Marr, H. Rex & S.H. Cohen. 2003. AmphotericinB: time for a new "gold standard." *Clin. Infect. Dis.* **37:** 415–425.

49. Wingard, J.R., P. Kubilis, L. Lee, *et al.* 1999. Clinical significance of nephrotoxicity in patients treated with amphotericin B for suspected or proven aspergillosis. *Clin. Infect. Dis.* **29:** 1402–1407.

50. Walsh, T.J., R.W. Finberg, C. Arndt, *et al.* 1999. Liposomal amphotericin B for empirical therapy in patients with persistent fever and neutropenia. National Institute of Allergy and Infectious Diseases Mycoses Study Group. *N. Engl. J. Med.* **340:** 764–771.

51. Johnson, B. & A. Kauffman. 2003. Voriconazole: a new triazole antifungal agent. *Clin. Infect. Dis.* **36:** 630–637.

52. Sambatakou, H., B. Dupont, H. Lode, *et al.* 2006. Voriconazole treatment for subacute invasive and chronic pulmonary aspergillosis. *Am. J. Med.* **119:** e517–e524; 527.

53. Herbrecht, R., D.W. Denning, T.F. Patterson, *et al.* 2002. Voriconazole versus amphotericin B for primary therapy of invasive aspergillosis. *N. Engl. J. Med.* **347:** 408–415.

54. Scherpbier, H.J., M.I. Hilhorst & T.W. Kuijpers. 2003. Liver failure in a child receiving highly active antiretroviral therapy and voriconazole. *Clin. Infect. Dis.* **37:** 828–830.

55. von Mach, M.A., J. Burhenne & L. Weilemann. 2006. Accumulation of the solvent vehicle sulphobutylether beta cyclodextrin sodium in critically ill patients treated with intravenous voriconazole under renal replacement therapy. *BMC Clin. Pharmacol.* **6:** 6–13.

56. Trof, R., A. Beishuizen, J. Debets-Ossenkopp, *et al.* 2007. Management of invasive pulmonary aspergillosis in non-neutropenic critically ill patients. *Intens. Care Med.* **33:** 1694–1703.

57. Park, W.B., N.H. Kim, K.H. Kim, *et al.* 2012. The effect of therapeutic drug monitoring on safety and efficacy of voriconazole in invasive fungal infections: a randomized controlled trial. *Clin. Infect. Dis.* **55:** 1080–1087.

58. Maertens, J., I. Raad, G. Petrikkos, *et al.* 2007. Efficacy and safety of caspofungin for treatment of invasive aspergillosis in patients refractory to or intolerant of conventional antifungal therapy. *Clin. Infect. Dis.* **39:** 1563–1571.

59. Aliff, B., P.G. Maslak, J.G. Jurcic, *et al.* 2003. Refractory *Aspergillus* pneumonia in patients with acute leukemia: successful therapy with combination caspofungin and liposomal amphotericin with combination caspofungin and liposomal amphotericin. *Cancer* **97:** 1025–1032.

60. Lewis, E. & D. Kontoyiannis. 2001. Rationale for combination antifungal therapy. *Pharmacotherapy* **21:** 149S–164S.

61. Steinbach, W.J. 2005. Combination antifungal therapy for invasive aspergillosis: utilizing new targeting strategies. *Curr. Drug Targets Infect. Disord.* **5:** 203–210.

62. Traunmüller, F., M. Popovic, K.H. Konz, *et al.* 2011. Efficacy and safety of current drug therapies for invasive aspergillosis. *Pharmacology* **88:** 213–224.

63. Kontoyiannis, D., R. Hachem, R.E. Lewis, *et al.* 2003. Efficacy and toxicity of caspofungin in combination with liposomal amphotericin B as primary or salvage treatment of invasive aspergillosi n patients with hematologic malignancies. *Cancer* **98:** 292–299.

64. Marr, A., M. Boeckh, R.A. Carter, *et al.* 2004. Combination antifungal therapy for invasive aspergillosis. *Clin. Infect. Dis.* **39:** 797–802.

65. Fluckiger, U., O. Marchetti, J. Bille, *et al.* 2006. Treatment options of invasive fungal infections in adult. *Swiss Med. Wkly.* **136:** 447–463.

66. Sutton, D.A., S.E. Sanche, S.G. Revankar, *et al.* 1999. In vitro amphotericin B resistance in clinical isolates of Aspergillus terreus, with a head-to-head comparison to voriconazole. *J. Clin. Microbiol.* **37:** 2343–2345.

67. Alcazar-Fuoli, L., E. Mellado, A. Alastruey-Izquierdo, *et al.* 2008. *Aspergillus* section *fumigati*: antifungal susceptibility patterns and sequence-based identification. *Antimicrob. Agents Chemother.* **52:** 1244–1251.

68. Pfaller, M.A., S.A. Messer, L. Boyken, *et al.* 2008. In vitro survey of triazole cross-resistance among more than 700 clinical isolates of Aspergillus species. *J. Clin. Microbiol.* **46:** 2568–2572.

69. Arendrup, M.C., G. Garcia-Effron, W. Buzina, *et al.* 2009. Breakthrough *Aspergillus* fumigatus and *Candida albicans* double infection during caspofungin treatment: laboratory characteristics and implication for susceptibility testing. *Antimicrob. Agents Chemother.* **53:** 1185–1193.

70. Andrew, H., Limper, S. Kenneth, *et al.* 2011. An Official American Thoracic Society Statement: treatment of fungal infections in adult pulmonary and critical care patients. *Am. J. Respir. Crit. Care Med.* **183:** 96–128.

71. Couaillier, J.F., A. Bernard, O. Casasnovas, *et al.* 2001. Increasing volume and changing characteristics of invasive pulmonary aspergillosis on sequential thoracic computed tomography scans in patients with neutropenia. *J. Clin. Oncol.* **19:** 253–259.

72. Boutboul, F., C. Alberti, T. Leblanc, *et al.* 2002. Invasive aspergillosis in allogeneic stem cell transplant recipients: increasing antigenemia is associated with progressive disease. *Clin. Infect. Dis.* **34:** 939–943.

73. Mullins, J., R. Harvey & A. Seaton. 1976. Sources and incidence of airborne Aspergillus fumigatus (Fres). *Clin. Allergy* **6:** 209–217.

74. Arnow, P.M., R.L. Andersen, P.D. Mainous & E.J. Smith. 1978. Pumonary aspergillosis during hospital renovation. *Am. Rev. Respir. Dis.* **118:** 49–53.

75. Sarubbi, F.A., Jr., H.B. Kopf, M.B. Wilson, *et al.* 1982. Increased recovery of Aspergillus flavus from respiratory specimens during hospital construction. *Am. Rev. Respir. Dis.* **125:** 33–38.

76. Anaissie, E.J., S.L. Stratton, R.C. Summerbell, *et al.* 2000. Pathogenic Aspergillus species recovered from a hospital water system: a three year prospective study. In: *Abstracts of the 40th Interscience, Conference on Antimicrobial Agents and Chemotherapy*. American Society for Microbiology, Washington, DC, 375.

77. Loo, V.G., C. Bertrand, C. Dixon, *et al.* 1996. Control of construction-associated nosocomial aspergillosis in an antiquated hematology unit. *Infect. Control Hosp. Epidemiol.* **17:** 360–364.

Ann. N.Y. Acad. Sci. ISSN 0077-8923

ANNALS OF THE NEW YORK ACADEMY OF SCIENCES

Issue: *Advances Against Aspergillosis*

Management of chronic pulmonary aspergillosis

Koichi Izumikawa, Masato Tashiro, and Shigeru Kohno

Department of Molecular Microbiology and Immunology, Nagasaki University Graduate School of Biomedical Sciences, Nagasaki, Japan

Address for correspondence: Koichi Izumikawa, M.D., Ph.D., Department of Molecular Microbiology and Immunology, Nagasaki University Graduate School of Biomedical Sciences, 1-7-1 Sakamoto, Nagasaki 852-8501, Japan. koizumik@nagasaki-u.ac.jp

Chronic pulmonary aspergillosis (CPA) is a relatively rare, slowly progressive pulmonary syndrome caused by *Aspergillus* spp. The scarcity of clinical evidence for its management is an important issue. Oral azoles are recommended as the primary treatment of CPA; however, the evidence for their effectiveness is insufficient. Azole-resistant *A. fumigatus* is rapidly increasing and becoming a serious concern. Because long-term administration of azoles is the mainstay of CPA, azole resistance may pose a serious threat. Furthermore, prolonged oral administration of azoles may lead to increased azole resistance in CPA patients. Therefore, alternative management strategies for CPA must be considered, and one option may involve the use of intravenous antifungals such as echinocandins and polyens. The utility of these antifungals, however, has not been well evaluated and remains controversial because the drugs are expensive and require patients to be admitted to the hospital for their use. New antifungal drugs with novel mechanisms of action are also needed.

Keywords: chronic pulmonary aspergillosis; azole resistance; azoles; echinocandins; polyens

Chronic pulmonary aspergillosis

The incidence of deep-seated fungal infections, along with increasing numbers of mild-to-severe immunocompromised hosts, is climbing due to advanced medical procedures and treatments. Aspergillosis is particularly important because of its high mortality and morbidity. There are more than 250 *Aspergillus* spp., including the common human pathogens *Aspergillus fumigatus, A. flavus, A. niger, A. terreus,* and *A. versicolor*.[1] Various types of aspergillosis, such as invasive chronic infections and allergic disease, can be caused by *Aspergillus* spp. Chronic pulmonary aspergillosis (CPA) is defined as a slowly progressive pulmonary syndrome. Compared to invasive pulmonary aspergillosis (IPA), the pathophysiology of CPA ranges widely from aspergilloma to chronic necrotizing pulmonary aspergillosis (CNPA), also known as subacute IPA, semi-invasive aspergilloma, symptomatic pulmonary aspergilloma, or *Aspergillus* pseudotuberculosis in the previous literature.[2–5] Recent consensus classification of CPA includes simple aspergilloma, chronic cavitary pulmonary aspergillosis (CCPA), and chronic fibrosing pulmonary aspergillosis (CFPA).[3] Figure 1 shows the summarized classification of pulmonary aspergillosis based on pathological and radiological features, time course, and innate immunity defect.[6–9] However, pathological samples may not be acquired in every CPA case and the time course is arbitrary. These subcategories of CPA may overlap and a clear distinction is difficult to make.

The clinical features of CPA have been characterized in several recent reports. CPA and aspergilloma usually occur in middle-aged to elderly patients with relatively lower body mass index.[10,11] Preexisting or residual cavities following mycobacterial infection are the most common dwelling sites of *Aspergillus*, and these cavitary lesions are mostly located in the upper lobes.[10–13] Other pulmonary diseases that result in cavity formation, such as emphysematous bullae, chronic obstructive pulmonary diseases, bronchiectasis, pulmonary

doi: 10.1111/j.1749-6632.2012.06758.x

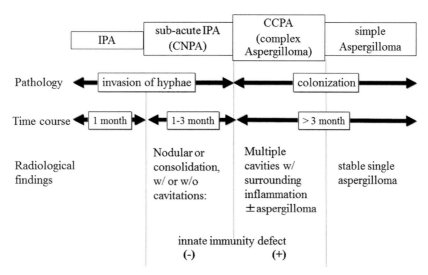

Figure 1. Classification of pulmonary aspergillosis based on pathological and radiological features, time course, and innate immunity defect.

cysts, and others, may be a cause of CPA and aspergilloma.[8,12]

CPA slowly causes lung destruction, for example, progressive cavitation, fibrosis, and pleural thickening, that is reflected in patients' radiological features. Single or multiple fungus balls may be seen in cavitary lesions. CPA patients often present with chronic pulmonary or systemic symptoms (usually for longer than a month), such as weight loss, low grade fever, productive cough, chronic sputum, hemosputum, or hemoptysis.[3,10,11,13] The worldwide prevalence of CPA is estimated at around three million, and its mortality is 50% over 5 years, but no large-scale epidemiological data are available.[11,14]

Serum *Aspergillus* precipitins tests that detect antibodies (IgG) to *Aspergillus* are usually positive[7,10,15,16] in CPA patients, including those with aspergilloma. Although serum *Aspergillus* galactomannan antigen ELISA tests have been approved for the diagnosis of IPA, with a sensitivity of 79% and a specificity of 86% in meta-analysis,[17] they are not considered useful for CPA diagnosis due to a lower positive rate.[10,16] Sputum, bronchoalveolar lavage cultures, and surgical biopsy specimens usually reveal the causative organism in pulmonary infections. However, because *Aspergilli* are ubiquitous in the environment, careful evaluation is required to confirm colonization or infection. Pathological examination by transbronchial, surgical, and computed tomography–guided biopsies are the confir-

matory methods of CPA diagnosis. However, they may not be applied in all cases owing to underlying diseases in the patients.

Management of CPA

The latest Infectious Diseases Society of America (IDSA) guidelines for the treatment of aspergillosis recommend oral voriconazole (VRCZ) or itraconazole (ITCZ) as the primary treatment for CNPA and CCPA, and surgical resection is recommended for simple aspergilloma cases.[9,18] Bronchial artery embolization (BAE) to occlude the causative vessels in patients with hemosputum or hemoptysis is another option for the treatment of CPA. However, BAE is only temporarily effective due to the presence of collateral vascular channels at the site of bleeding.

Only one randomized study for the management of CPA has been conducted,[13] and some case series reports are currently available. Additionally, the following issues have not been standardized or even described in the IDSA guidelines: timing of the initiation of treatment, duration of treatment, and timing of the discontinuation of treatment for CPA.

The efficacy of oral azoles for CPA is summarized in Table 1.[3,11,19–29] From the late 1980s to today the reported efficacy of oral ITCZ is somewhat variable, with a range of 30–93% duration administration approximately 4–12 months, and adverse effects in 16–33% of patients. In the last decade, four reports

Table 1. The efficacy of oral antifungals to chronic pulmonary aspergillosis

Antifungals	CPA	Response rate (%)	Route	Study design	Year reported	Reference
Itraconazole	CNPA+aspergilloma	60–66	Oral	Case series	1988	19
Itraconazole	CNPA+aspergilloma	71–93	Oral	Case series	1990	20
Itraconazole	Aspergilloma	30	Oral	Case series	1991	21
Itraconazole	CNPA	67	Oral	Case series	1996	22
Itraconazole	CNPA	67	Oral	Case series	1997	23
Itraconazole	Aspergilloma	63	Oral	Case series	1997	24
Itraconazole	CPA	71	Oral	Case series	2003	3
Itraconazole	CNPA	38	Oral	Case series	2009	11
Voriconazole	CCPA	64	Oral	Case series	2006	25
Voriconazole	CNPA+CCPA	50–80	Oral	Case series	2006	26
Voriconazole	CNPA+CCPA	70	Oral	Case series	2007	27
Voriconazole	CNPA+CCPA	30–43	Oral	Case series	2012	28
Posaconazole	CPA	46–61	Oral	Case series	2010	29

CNPA, chronic necrotizing pulmonary aspergillosis; CPA, chronic pulmonary aspergillosis; CCPA, chronic cavitary pulmonary aspergillosis.

on the efficacy of oral VRCZ have been published: response rates ranged from 30% to 80% following several months of administration, and the frequency of adverse effects that resulted in the discontinuation of therapy ranged from 9% to 27%.[25–28] The efficacy of ITCZ and VRCZ in most of these studies was assessed by clinical, radiological, and mycological improvement at the end of treatment or at regular intervals, regardless of whether there was a partial or complete response. Figure 2 indicates the radiographs of CPA patient with "good response" to long-term azole treatment. Recent case series reports regarding posaconazole showed a response of 61% at 6 months and 46% at 12 months, which was similar to the data for other oral ITCZ or VRCZ treatments.[29]

Table 2. The efficacy of injectable antifungals to chronic pulmonary aspergillosis

Antifungals	CPA	Response rate (%)	Route	Study design	Year reported	Reference
Micafungin	CNPA+ aspergilloma	55–67	Intravenous	Case series	2004	31
Micafungin	CPA	78	Intravenous	Case series	2007	30
Micafungin versus voriconazole	CPA	60 (micafungin) versus 53 (voriconazole)	Intravenous	Randomized controlled study	2010	10
Micafungin	CPA	68	Intravenous	Case series	2011	13
Caspofungin versus micafungin	CPA	30 (caspofungin) versus 33 (micafungin)	Intravenous	Randomized controlled study	2012	Paper submitted
Liposomal amphotericin B versus voriconazole	CPA	Currently underway (150 cases in trial)	Intravenous	Randomized controlled study		Under analysis

CNPA, chronic necrotizing pulmonary aspergillosis; CPA, chronic pulmonary aspergillosis.

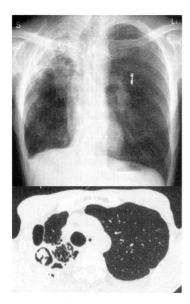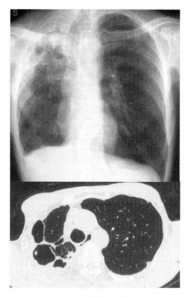

before treatment

8 months after
azole treatment

Figure 2. A patient with chronic pulmonary aspergillosis showed good response to azole. Multiple cavitary lesions with fungus ball and pericavity infiltrations were presented in right upper lung field before treatment. Fungus ball and infiltrations were diminished and improved, respectively, after eight months of azole treatment.

There are limited published data on the use of injectable antifungal agents such as echinocandins, azoles, and amphotericin B in the treatment of CPA (Table 2). Several series of studies evaluating the efficacy of intravenous micafungin (MCFG) in CPA were exclusively reported by Kohno and Izumikawa. The first two case series, including one clinical trial for marketing approval conducted in Japan, indicated that the success rates of treatment were 12/22 cases (duration of treatment; 11–57 days) and 7/9 cases (duration of treatment; 29–96 days).[30,31] Another prospective observational study to evaluate MCFG was also conducted and the overall clinical efficacy rate was 68.4% (26/38 patients), which is comparable to the results obtained in a previous small study.[31]

A randomized controlled study of CPA therapy comparing intravenous VRCZ and intravenous MCFG reported a favorable response rate with both MCFG (60%) and VRCZ (53%) with no significant difference.[10] However, fewer adverse events occurred in the MCFG than that in the VRCZ group (26.4% vs. 61.1%, $P = 0.0004$) in the safety evaluation. MCFG and VRCZ were equally effective, but MCGF was significantly safer as an initial treatment for CPA patients who required immediate hospital admission.[10] We conducted this series of studies on MCFG in Japan in patients who required immediate treatment. The treatment duration was a maximum of four weeks in almost all studies, and the assessment was performed at the end of treatment. Therefore, outcomes or prognosis a few months or years after the treatment were not assessed.

Most recently, another randomized controlled study comparing intravenous MCFG and caspofungin (CPFG) was conducted as a part of an approval study of CPFG in Japan. Overall response of CPFG and MCFG in CPA patients was 46.7% (14/30 cases) and 42.4% (14/33), respectively. There was no statistical difference in the safety profile between both groups (paper submitted). Studies assessing the utility of anidulafungin for CPA are not currently available. Another randomized controlled study comparing intravenous liposomal amphotericin B and intravenous VRCZ is currently underway.

Many CPA patients require maintenance treatments of oral azoles. Importantly, comparison of oral azoles during long-term treatment is particularly required. However, due to the complexity of its pathogenesis and the underlying diseases of CPA, conducting studies while establishing good study design and ensuring the efficacy of the assessment

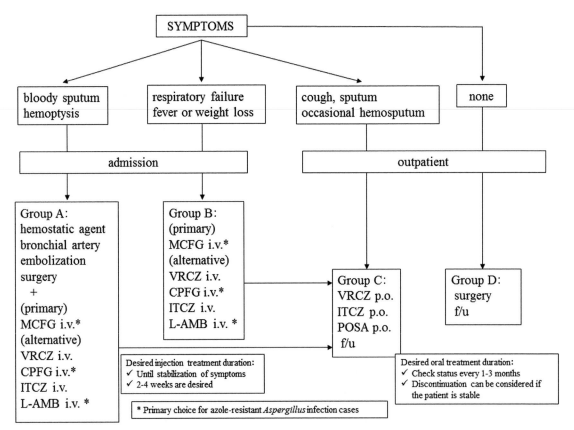

Figure 3. Proposal for the management of CPA in Japan based on the severity of symptoms and status of the patient. The intravenous administration of antifungals may be desired for two to four weeks, followed by oral antifungals. MCFG, micafungin; VRCZ, voriconazole; CPFG, caspofungin; ITCZ, itraconazole; L-AMB, liposomal amphotericin B; f/u, follow-up.

methods for long-term oral azole administration may be difficult.

Although evidence regarding the treatment of CPA is limited, the accumulation of data to date highlights the need for more detailed recommendations for CPA management. The IDSA guidelines indicate oral azoles as a primary treatment and other injectable antifungals as alternative treatments, but it is apparent that some patients require hospital admission and immediate intravenous administration of antifungals. Figure 3 shows a proposal for the management of CPA in Japan based on a classification of the severity of symptoms and status. Since the healthcare system varies in each country, our proposal may not be adaptable in other countries. The intravenous administration of antifungals may be desired for two to four weeks, followed by oral antifungals. There are, however, issues such as the costs of antifungals and admission, tolerance of long-term administration, and

scarcity of evidence. Most importantly, the presence of azole-resistant *Aspergillus* is becoming a major threat to CPA management. Because oral azole is the mainstay of CPA treatment, particularly maintenance treatment, azole resistance may eliminate the candidates for long-term administration of antifungals. Additionally, very few new antifungals in both injectable and oral formulations are currently being developed, which is a serious issue not only for the management of CPA but also for all other *Aspergillus* infections.

Impact of azole-resistant *Aspergillus* infection and relation to CPA

Drug resistance in *A. fumigatus* has not been a major research focus in the last decade, with only a few studies performed after the first report by Denning *et al.* of itraconazole resistance among CPA patients in 1997.[32] Recent advances in standardizing susceptibility testing by the Clinical and Laboratory

Table 3. Major epidemiological studies of azole resistance in *A. fumigatus* after 2008

Reported by	Year reported	Number of tested isolates	Number of azole resistant isolates (azole tested in each study)	Rate of azole resistant strains (%)	Region	Reference
Guinea *et al.*	2008	374	0 (voriconazole)	0.0	Spain	39
Snelders *et al.*	2008	1912	32 (itraconazole)	1.7	the Netherlands	37
Rodriguez-Tudela *et al.*	2008	393	32 (itraconazole)	8.1	Spain, the Netherlands, UK, France	35
Espinel-Ingroff *et al.*	2008	292	1 (voriconazole)	0.3	North America	40
Howard *et al.*	2009	519	34 (itraconazole)	0.6	UK	38
Pfaller *et al.*	2009	637	43 (itraconazole)	6.8	Worldwide	41
Baddley *et al.*	2009	181	1 (itraconazole)	0.6	North America	1
Amorium *et al.*	2010	159	1 (posaconazole)	0.6	Portugal	42
Bueid *et al.*	2010	230	62 (itraconazole)	27.0	UK	43
Lockhart *et al.*	2011	497	29 (itraconazole)	5.8	Worldwide*	44
Chowdhary *et al.*	2012	103	2 (itraconazole)	1.9	India	45
Tashiro *et al.*	2012	196	14 (itraconazole)	7.1	Japan	46

*Itraconazole-resistant strains were detected mainly from China.

Standards Institute (M38-A2) and the European Committee on Antimicrobial Susceptibility Testing,[33,34] together with the establishment of interpretative cutoffs[35] and breakpoints,[36] have greatly enhanced clinical and basic research. Two major epidemiological studies from the Netherlands and the United Kingdom were published in 2008 and 2009, respectively.[37,38] Extensive epidemiological data have accumulated in the last few years highlighting that azole-resistant *A. fumigatus* isolates have been recognized in many countries.[1,35,37–46] Table 3 indicates the results of recent epidemiological studies on azole resistance in *A. fumigatus* since 2008. The prevalence of azole resistance is between approximately 0.3 and 28%. However, the definition of resistance varies in each study listed in Table 3, and caution is required when interpreting the data.

Resistance mechanisms against antifungal drugs, including azoles, are described elsewhere.[47] In *A. fumigatus*, modification and overexpression of target enzymes due to mutations and activation of drug efflux pumps are major resistance mechanisms. The primary azole resistance mechanism is mutation of the target protein 14 alpha-demethylase encoded by *cyp51*A, and numerous mutation hotspots have been confirmed.[38] Two possible hypotheses for the evolution of azole resistance have been proposed to date.

Several molecular epidemiological analyses have indicated that azole resistance is acquired in infecting strains within the human lung, rather than by inhalation of strains that acquired resistance in the environment before infection.[48,49] Importantly, many azole-resistant strains were isolated from CPA patients with aspergilloma, many of whom had been exposed to azoles for an extended duration (1–30 months).[48,50] We have also recently described a detailed relationship between azole usage (mainly ITCZ) and azole MICs for *A. fumigatus* isolated in our facility. Briefly, we confirmed the presence of acquired resistance to ITCZ and POSA in a patient with CPA after consecutive oral ITCZ therapy in Japan.[51] Our data also supported the hypothesis on resistance stated above, and although long-term administration of oral azoles is the mainstay of treatment of CPA, it may potentially induce resistance.

In contrast, Verweij *et al.* suggested the possibility that environmental fungicides used for agricultural purpose may induce mutations in *cyp51*A, and these environmental strains become a source of *Aspergillus* infection in humans.[52,53] This hypothesis

Table 4. Summary of drug resistance issues of azoles and other antifungals

Antifungals	Administration route	Existence of treatment failure by drug resistance	Possibility of acquired resistance by treatment
Itraconazole	Oral or injection	Reported[38]	Reported[51]
Voriconazole	Oral or injection	Reported[38]	Reported[55,56]
Posaconazole	Oral only	Reported[54]	Reported[51]
Micafungin	Injection only	Not reported	Not reported
Caspofungin	Injection only	Not reported	Not reported
Amphotericin B	Injection only	Not reported	Not reported

is supported by the fact that a single resistance mechanism (TR/L98H; tandem repeat of promoter region and mutation of L98H in *cyp51A*) is predominant in both clinical and environment isolates in the Netherlands, where extremely high amounts of sterol demethylation inhibitor fungicides are used.[53] Interestingly, these sterol demethylation inhibitor fungicides cause cross-resistance to medical azoles, and seven of approximately 30 agricultural fungicides showed cross-resistance with medical azoles.[14]

Although the increase in azole-resistant *A. fumigatus* has become an important topic in the field, the evidence that azole resistant isolates are associated with poorer outcome compared to susceptible strains is very limited to date. Only a few case series and case reports are available, and Table 4 summarizes the drug resistance issues of azoles and other antifungals.[38,51,54–56] Hence, the impact of azole-resistant *A. fumigatus* in CPA cases remains controversial. However, it is apparent that the only other potent antifungals for CPA are intravenous antifungals such as echinocandins and polyens (Fig. 3).

Future direction of management of CPA with azole-resistant *A. fumigatus*

It is obvious that further understanding of how *A. fumigatus* acquires drug resistance is required; in addition, a global network system to monitor the epidemiology data and seek novel drug-resistant mechanisms needs to be established. New antifungal drugs in either oral or intravenous forms with novel mechanisms of action are also certainly required. Other studies to optimize azole regimen pharmacokinetics and pharmacodynamics and therapeutic

drug monitoring data to prevent mutation of *cyp51A* and to acquire maximum efficacy are also needed.

Conclusions

Studies regarding CPA management remain limited. A new detailed management scheme based on the severity of CPA is proposed, and the application of injectable and oral antifungals may be considered in each case. The emergence of azole-resistant *Aspergillus* is becoming a major issue in the treatment of aspergillosis. CPA may have its origin in the production of azole-resistant strains during treatment, which has limited the options for antifungal selection. It is therefore necessary to eliminate the continued development of drug-resistance that leads to an increasingly poor prognosis. Further investigations are required to understand the evolution of azole resistance in *A. fumigatus*.

Conflicts of interest

K.I. has received honoraria from Pfizer Japan, Inc., Astellas Pharma Inc., Dainippon Sumitomo Pharma Co., Ltd., and Merck & Co., Inc. S.K. has received research grants and honoraria from Pfizer Japan, Inc., Astellas Pharma Inc., Dainippon Sumitomo Pharma Co., Ltd., and Merck & Co., Inc.

References

1. Baddley, J.W. *et al.* 2009. Patterns of susceptibility of *Aspergillus* isolates recovered from patients enrolled in the Transplant-Associated Infection Surveillance Network. *J. Clin. Microbiol.* **47:** 3271–3275.
2. Binder, R.E. *et al.* 1982. Chronic necrotizing pulmonary aspergillosis: a discrete clinical entity. *Medicine* **61:** 109–124.
3. Denning, D.W. *et al.* 2003. Chronic cavitary and fibrosing pulmonary and pleural aspergillosis: case series, proposed nomenclature change, and review. *Clin. Infect. Dis.* **37**(Suppl 3): S265–S280.

4. Gefter, W.B. *et al.* 1981. "Semi-invasive" pulmonary aspergillosis: a new look at the spectrum of aspergillus infections of the lung. *Radiology* **140:** 313–321.

5. Yousem, S.A. 1997. The histological spectrum of chronic necrotizing forms of pulmonary aspergillosis. *Hum. Pathol.* **28:** 650–656.

6. Denning, D.W. 2008. Aspergillosis. McGraw-Hill. New York.

7. Denning, D.W. 2001. Chronic forms of pulmonary aspergillosis. *Clin. Microb. Infect.* **7**(Suppl 2): 25–31.

8. Hope, W.W., T.J. Walsh & D.W. Denning. 2005. The invasive and saprophytic syndromes due to *Aspergillus* spp. *Med. Mycol.* **43**(Suppl 1): S207–S238.

9. Walsh, T.J. *et al.* 2008. Treatment of aspergillosis: clinical practice guidelines of the Infectious Diseases Society of America. *Clin. Infect. Dis.* **46:** 327–360.

10. Kohno, S. *et al.* 2010. Intravenous micafungin versus voriconazole for chronic pulmonary aspergillosis: a multicenter trial in Japan. *J. Infect.* **61:** 410–418.

11. Nam, H.S. *et al.* 2010. Clinical characteristics and treatment outcomes of chronic necrotizing pulmonary aspergillosis: a review of 43 cases. *Int. J. Infect. Dis.* **14:** e479–e482.

12. Smith, N.L. & D.W. Denning. 2011. Underlying conditions in chronic pulmonary aspergillosis including simple aspergilloma. *Eur. Respir. J.* **37:** 865–872.

13. Kohno, S. *et al.* 2011. Clinical efficacy and safety of micafungin in Japanese patients with chronic pulmonary aspergillosis: a prospective observational study. *Med. Mycol.* **49:** 688–693.

14. Denning, D.W. & D.S. Perlin. 2011. Azole resistance in *Aspergillus*: a growing public health menace. *Fut. Microbiol.* **6:** 1229–1232.

15. Zmeili, O.S. & A.O. Soubani. 2007. Pulmonary aspergillosis: a clinical update. *QJM* **100:** 317–334.

16. Kitasato, Y. *et al.* 2009. Comparison of *Aspergillus* galactomannan antigen testing with a new cut-off index and *Aspergillus* precipitating antibody testing for the diagnosis of chronic pulmonary aspergillosis. *Respirology* **14:** 701–708.

17. Pfeiffer, C.D., J.P. Fine & N. Safdar. 2006. Diagnosis of invasive aspergillosis using a galactomannan assay: a meta-analysis. *Clin. Infect. Dis.* **42:** 1417–1427.

18. Regnard, J.F. *et al.* 2000. Aspergilloma: a series of 89 surgical cases. *Ann. Thorac. Surg.* **69:** 898–903.

19. De Beule, K. *et al.* 1988. The treatment of aspergillosis and aspergilloma with itraconazole, clinical results of an open international study (1982–1987). *Mycoses* **31:** 476–485.

20. Dupont, B. 1990. Itraconazole therapy in aspergillosis: study in 49 patients. *J. Am. Acad. Dermatol.* **23:** 607–614.

21. Campbell, J.H. *et al.* 1991. Treatment of pulmonary aspergilloma with itraconazole. *Thorax* **46:** 839–841.

22. Caras, W.E. & J.L. Pluss. 1996. Chronic necrotizing pulmonary aspergillosis: pathologic outcome after itraconazole therapy. *Mayo Clin. Proc.* **71:** 25–30.

23. Saraceno, J.L. *et al.* 1997. Chronic necrotizing pulmonary aspergillosis: approach to management. *Chest* **112:** 541–548.

24. Tsubura, E. 1997. Multicenter clinical trial of itraconazole in the treatment of pulmonary aspergilloma. Pulmonary Aspergilloma Study Group. *Kekkaku* **72:** 557–564.

25. Jain, L.R. & D.W. Denning. 2006. The efficacy and tolerability of voriconazole in the treatment of chronic cavitary pulmonary aspergillosis. *J. Infect.* **52:** e133–e137.

26. Sambatakou, H. *et al.* 2006. Voriconazole treatment for subacute invasive and chronic pulmonary aspergillosis. *Am. J. Med.* **119:** e517–e524.

27. Camuset, J. *et al.* 2007. Treatment of chronic pulmonary aspergillosis by voriconazole in nonimmunocompromised patients. *Chest* **131:** 1435–1441.

28. Cadranel, J. *et al.* 2012. Voriconazole for chronic pulmonary aspergillosis: a prospective multicenter trial. *Eur. J. Clin. Microb. Infect. Dis.* Published online July 11.

29. Felton, T.W. *et al.* 2010. Efficacy and safety of posaconazole for chronic pulmonary aspergillosis. *Clin. Infect. Dis.* **51:** 1383–1391.

30. Izumikawa, K. *et al.* 2007. Clinical efficacy of micafungin for chronic pulmonary aspergillosis. *Med. Mycol.* **45:** 273–278.

31. Kohno, S. *et al.* 2004. A multicenter, open-label clinical study of micafungin (FK463) in the treatment of deep-seated mycosis in Japan. *Scand. J. Infect. Dis.* **36:** 372–379.

32. Denning, D.W. *et al.* 1997. Itraconazole resistance in *Aspergillus fumigatus*. *Antimicrob. Agents Chemother.* **41:** 1364–1368.

33. EUCAST technical note on the method for the determination of broth dilution minimum inhibitory concentrations of antifungal agents for conidia-forming moulds. 2008. *Clin. Microbiol. Infect.* **14:** 982–984.

34. Reference method for broth dilution antifungal susceptibility testing of filamentous fungi. Approved standard, 2nd ed. Document M38-A2. 2008. Clinical Laboratory Standards Institute, Wayne, PA.

35. Rodriguez-Tudela, J.L. *et al.* 2008. Epidemiological cutoffs and cross-resistance to azole drugs in *Aspergillus fumigatus*. *Antimicrob. Agents Chemother.* **52:** 2468–2472.

36. Verweij, P.E. *et al.* 2009. Azole-resistance in *Aspergillus*: proposed nomenclature and breakpoints. *Drug Resist. Updat.* **12:** 141–147.

37. Snelders, E. *et al.* 2008. Emergence of azole resistance in *Aspergillus fumigatus* and spread of a single resistance mechanism. *PLoS Med.* **5:** e219.

38. Howard, S.J. *et al.* 2009. Frequency and evolution of Azole resistance in *Aspergillus fumigatus* associated with treatment failure. *Emerg. Infect. Dis.* **15:** 1068–1076.

39. Guinea, J. *et al.* 2008. Clinical isolates of *Aspergillus* species remain fully susceptible to voriconazole in the post-voriconazole era. *Antimicrob. Agents Chemother.* **52:** 3444–3446.

40. Espinel-Ingroff, A. *et al.* 2008. Activities of voriconazole, itraconazole and amphotericin B in vitro against 590 moulds from 323 patients in the voriconazole phase III clinical studies. *J. Antimicrob. Chemother.* **61:** 616–620.

41. Pfaller, M.A. *et al.* 2009. Wild-type MIC distribution and epidemiological cutoff values for *Aspergillus fumigatus* and three triazoles as determined by the Clinical and Laboratory Standards Institute broth microdilution methods. *J. Clin. Microbiol.* **47:** 3142–3146.

42. Amorim, A., L. Guedes-Vaz & R. Araujo. 2010. Susceptibility to five antifungals of *Aspergillus fumigatus* strains isolated from chronically colonised cystic fibrosis patients receiving azole therapy. *Int. J. Antimicrob. Agents* **35:** 396–399.

43. Bueid, A. *et al.* 2010. Azole antifungal resistance in *Aspergillus fumigatus*: 2008 and 2009. *J. Antimicrob. Chemother.* **65:** 2116–2118.

44. Lockhart, S.R. *et al.* 2011. Azole resistance in *Aspergillus fumigatus* isolates from the ARTEMIS global surveillance study is primarily due to the TR/L98H mutation in the *cyp51A* gene. *Antimicrob. Agents Chemother.* **55:** 4465–4468.

45. Chowdhary, A. *et al.* 2012. Isolation of multiple-triazole-resistant *Aspergillus fumigatus* strains carrying the TR/L98H mutations in the *cyp51A* gene in India. *J. Antimicrob. Chemother.* **67:** 362–366.

46. Tashiro, M. *et al.* 2012. Antifungal susceptibilities of *Aspergillus fumigatus* clinical isolates in Nagasaki, Japan. *Antimicrob. Agents Chemother.* **56:** 584–587.

47. Cowen, L.E. 2008. The evolution of fungal drug resistance: modulating the trajectory from genotype to phenotype. *Nature Rev. Microbiol.* **6:** 187–198.

48. Chen, J. *et al.* 2005. Mutations in the *cyp51A* gene and susceptibility to itraconazole in *Aspergillus fumigatus* serially isolated from a patient with lung aspergilloma. *J. Antimicrob. Chemother.* **55:** 31–37.

49. Howard, S.J., A.C. Pasqualotto & D.W. Denning. 2010. Azole resistance in allergic bronchopulmonary aspergillosis and *Aspergillus* bronchitis. *Clin. Microb. Infect.* **16:** 683–688.

50. Bellete, B. *et al.* 2010. Acquired resistance to voriconazole and itraconazole in a patient with pulmonary aspergilloma. *Med. Mycol.* **48:** 197–200.

51. Tashiro, M. *et al.* 2012. Correlation between triazole treatment history and susceptibility in clinically isolated *Aspergillus fumigatus*. *Antimicrob. Agents Chemother.* **56:** 4870–4875.

52. Verweij, P.E. *et al.* 2009. Azole resistance in *Aspergillus fumigatus*: a side-effect of environmental fungicide use? *Lancet Infect. Dis.* **9:** 789–795.

53. Snelders, E. *et al.* 2009. Possible environmental origin of resistance of *Aspergillus fumigatus* to medical triazoles. *Appl. Environ. Microbiol.* **75:** 4053–4057.

54. Kuipers, S. *et al.* 2011. Failure of posaconazole therapy in a renal transplant patient with invasive aspergillosis due to *Aspergillus fumigatus* with attenuated susceptibility to posaconazole. *Antimicrob. Agents Chemother.* **55:** 3564–3566.

55. Thors, V.S. *et al.* 2011. Pulmonary aspergillosis caused by a pan-azole-resistant *Aspergillus fumigatus* in a 10-year-old boy. *Ped. Infect. Dis. J.* **30:** 268–270.

56. Arendrup, M.C. *et al.* 2010. Development of azole resistance in *Aspergillus fumigatus* during azole therapy associated with change in virulence. *PLoS One* **5:** e10080.

Ann. N.Y. Acad. Sci. ISSN 0077-8923

ANNALS OF THE NEW YORK ACADEMY OF SCIENCES
Issue: *Advances Against Aspergillosis*

The use of biological agents for the treatment of fungal asthma and allergic bronchopulmonary aspergillosis

Richard B. Moss

Department of Pediatrics, Stanford University School of Medicine, Stanford, California

Address for correspondence: Richard B. Moss, M.D., Center for Excellence in Pulmonary Biology, 770, Welch Road Suite 350, Palo Alto, CA 94304-5882. rmoss@stanford.edu

Allergic bronchopulmonary aspergillosis (ABPA) is a virulent manifestation of the Th2 asthma endotype that includes asthma with fungal sensitization, raising the feasibility of biological therapies targeting Th2 pathway molecules or cells. The first molecule amenable to clinical intervention with a biological was IgE. Omalizumab, a humanized monoclonal antibody (Mab), targets the same epitope on the IgE CH3 region that binds to and crosslinks high-affinity receptors on mast cells and basophils, thereby initiating the allergic inflammatory cascade. Omalizumab is licensed for allergic asthma and has been beneficial in uncontrolled studies of ABPA, reducing exacerbations and steroid requirements. Trials of several Mabs directed against the Th2 cytokine IL-5 show clinical benefit in patients with a severe refractory eosinophilic asthma phenotype, while a Mab against IL-13 is effective in asthma patients with a Th2-high endotype. Immunodulation is also feasible with small molecule biologicals, such as antisense oligodeoxynucleotides and cholecalciferol. Controlled trials of Th2-inhibiting biologicals in patients with ABPA and severe asthma with fungal sensitization appear warranted.

Keywords: asthma; ABPA; phenotype; endotype; cytokine; omalizumab

Asthma is a chronic inflammatory disease of the airways characterized clinically by intermittent episodes of wheezy shortness of breath, chest tightness, and cough. Pulmonary function tests show bronchoconstriction that is at least partly reversible with acute bronchodilator administration. The airways of people with asthma are hyperresponsive to bronchoconstrictive stimuli. Asthma is one of the most prevalent chronic diseases of humankind, with an estimated 300 million cases worldwide, including 26 million Americans (35% of whom are below 18 years of age). The social cost of asthma is staggering: about $20 billion in the United States in 2010, including over $5 billion in hospital costs, not to mention missed school or work and restricted activity. Acute asthma can be fatal. It is estimated that over half of the total costs of asthma are incurred by the 10–20% of asthmatics with severe disease. Depending on age, between half and three quarters of asthmatics are thought to have an allergic contribution or cause of their disease.[1]

Fungi have long been known to be among the causative agents of acute asthma in atopic patients with fungal sensitization. Fungal exposure has been linked to loss of asthma control, and more recently as a cause of asthma onset in both children and adults. A wide variety of fungi have been implicated, but the most common agents are several Ascomycota, including *Alternaria, Aspergillus, Penicillium*, and *Cladosporium* spp.[2] Recently the connection between fungal exposure, sensitization, and increased severity of asthma has become clearer.[3,4] *Aspergillus fumigatus* in particular has been associated with more severe asthma,[5] with pooled prevalence of sensitization in 28% of asthmatics seen in specialty clinics.[6] Sensitization to *A. fumigatus* is associated with lower lung function in asthma,[7] and antifungal therapy improves symptoms in severe asthmatics with fungal sensitization (SAFS).[8]

Allergic bronchopulmonary aspergillosis (ABPA) is the most severe manifestation of fungal asthma, occurring in ~2% of asthmatics, and is also a major complication in cystic fibrosis.[9] In addition to fungal sensitization (to *A. fumigatus* in >90% of cases), ABPA is characterized by colonization and fungal growth in the airways, a florid allergic and

doi: 10.1111/j.1749-6632.2012.06810.x

mixed granulocytic local inflammatory response, and progressive structural destruction of the airways (bronchiectasis and fibrosis) unless treated. Systemic corticosteroids and azoles are mainstays of ABPA therapy but treatment is impeded by difficulties in diagnosis, side effects of treatment, and the chronic relapsing natural history of this disease. The global burden of ABPA is estimated at ~4 million cases, with >500,000 in the United States.[10] Given the current limitations of conventional therapy for fungal asthma and ABPA and the severity of the asthma seen in this group, we propose that new therapies are needed to improve control and outcomes, with a significant role for emerging biological drugs. Before discussing these it is important to frame the approach in the context of our evolving understanding of asthma.

Asthma phenotypes and endotypes

Clinicians have long been used to characterizing people with asthma according to whether they had associated allergies and were sensitized to common aeroallergens, such as pollens, dust mite, animal danders, cockroach, and fungi. The distinction of allergic asthma from nonallergic (or intrinsic) was given a mechanistic underpinning by the elucidation of a CD4[+] T cell Th1/Th2 cytokine differentiation dichotomy in murine models, which was soon successfully applied in clinical asthma to show that a substantial element of Th2 polarization is present in the airways of many asthma patients.[11] However, in the 1990s further research revealed that this simple Th1/Th2 dichotomy was inadequate to encompass and adequately explain the broad range of clinical asthma and associated adaptive immune responses.[12] In the last decade, therefore, there has emerged a major effort to reassess asthma and define subgroups from the viewpoint of clinically observable characteristics, or phenotypes (Table 1). Some have gone so far as to plea to abandon the term *asthma* altogether, as it seems more of a conceptual hindrance than a diagnostic or therapeutic aide.[20] In a 2006 review, Wenzel extended the clinical view of phenotypes in persistent adult asthma to include categories based on clinical characteristics, triggers and predominant inflammatory granulocytic cell type.[16] Similar distinctions have been made in childhood asthma. Notably, phenotypes based on simple criteria involving one or few clinical features have been criticized as "one dimensional," and a more

Table 1. Asthma phenotypes, as organized by clinical presentation/features, precipitating factors, and character of cellular inflammation[13–19]

Clinical presentation/features
  Severity
  Hereditary, early onset allergic asthma
  Poorly reversible, very severe, neutrophilic asthma
  Late onset eosinophilic asthma
  Late onset, symptom dominant, obese minimal
     inflammation
  Exacerbation proneness
  Chronic airflow restriction
  Poorly steroid responsive
  Age at onset
  Pediatric
  Adult
  Cluster analysis[13–15]
  Early onset atopic (mild–moderate/severe)
  Late onset obese female noneosinophilic
  Early onset noneosinophilic
  Late onset eosinophilic
  Reduced lung function (more/less reversible)
Precipitating factors
  Nonsteroid anti-inflammatory agents
  Environmental allergens
  Occupational allergens or irritants
  Menses
  Exercise induced
  Ozone
  Cigarette smoke
  Diesel particles
  Infection
  Aspirin
  Cold air
  Obesity related
Character of cellular inflammation
  Eosinophilic
  Neutrophilic
  Mixed
  Pauci-granulocytic

sophisticated multidimensional approach using statistical cluster analysis was proposed and has recently been applied.[21] Using this methodology Haldar *et al.* identified three clusters in mild–moderate asthma and four clusters in severe asthma.[13] Moore *et al.*, examining asthma over the entire severity spectrum, identified five clusters in which atopy was present in >75% of the total cases and severe asthma was present in about a third.[14] McGrath *et al.* reported that about half of mild–moderate asthmatics do not have a persistent eosinophilic phenotype.[22] The data suggest that increasing asthma severity is associated with allergic sensitization. This conforms well with studies demonstrating the association of fungal sensitization with increasing asthma severity.[4,5,7,8,23]

The most recent development in this effort to dissect asthma into meaningful subgroups has been the identification of distinct pathophysiologic mechanisms underlying the emergence of particular asthma phenotypes, first proposed by Anderson in 2008 with the introduction of the term *endotype*.[24] A consensus report from the European and American Academies of Allergy cited six examples of asthma endotypes (aspirin-sensitive, allergic bronchopulmonary mycoses [ABPM], adult allergic, early-onset allergic, severe late-onset hypereosinophilic, and asthma in cross-country skiers).[17] In this conception, adult allergic asthma and ABPM endotypes, for example, exist within at least two phenotypes, eosinophilic asthma and exacerbation-prone asthma. The complex genetic, molecular, and cellular basis of the endotypic heterogeneity of asthma is being slowly, but surely, elucidated.[18] The Th2 pathway (Fig. 1) is perhaps the most well-studied and best understood of these asthma endotypes, and particularly useful in severe asthma.[19]

The usefulness of this endotypic approach to discernable asthma phenotypes is that it begins to allow rational targeted therapeutic interventions to be defined not only theoretically by disease mechanism but also practically as selection criteria in clinical trials using biomarkers associated with a particular endotype.[25,26] This approach has now begun to be applied, as will be discussed later. It can be seen that fungal asthma and ABPA (as *A. fumigatus* is by far the most common cause of ABPM, accounting for well over 90% of cases) are Th2 endotypes based on extensive examination of their pathophysiologic features, including demonstrable involvement of Th2 cytokines, IgE, eosinophils, and basophils.[9,10,23,27] The Th2 pathway thus retains great explanatory power and therapeutic potential for a substantial number of people with asthma, especially those with fungal asthma and ABPA.[28]

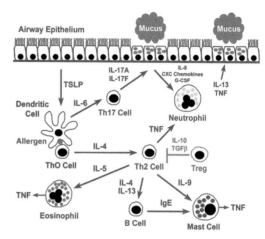

Figure 1. Cells and cell products of the Th2 immunoinflammatory asthma pathway. Respiratory epithelial cells (activated by interactions with products such as fungal pathogen-associated molecular patterns, allergens and proteases) secrete a variety of innate immune molecules including thymic stromal lymphopoeitin (TSLP), which in turn activate pulmonary dendritic cells to induce differentiation of naive CD4+ T cells (Th0) into Th2 cells. Th2 cells are polarized to secretion of a discrete set of cytokines, including IL-4, IL-5, IL-9, IL-13, and tumor necrosis factor (TNF-α). Th2 cytokines orchestrate airway eosinophil and mast cell recruitment, B cell production of immunoglobulin E, mucus secretion, and mast cell (and basophil) priming for allergic response to encountered allergen. Dendritic cell–derived IL-6 promotes differentiation of Th2 and Th17 cells. IL-17 from Th17 cells promotes airway neutrophil entry by inducing epithelial cell production of chemoattractant cytokines, such as CXCL8 (interleukin-8) and granulocyte colony-stimulating factor (G-CSF), a survival and proliferation factor, by bronchial epithelial cells. IL-17 also induces epithelial mucus secretion. The process is downregulated by regulatory T cells (T_reg cells) secreting IL-10 and transforming growth factor-β (TGF-β). Reprinted with permission from Ref. 50.

Anti-IgE

Anti-IgE, or omalizumab, was the first and currently the only biological licensed to treat asthma. It is based on the differentiation of allergic from nonallergic asthma, at the distal end of the Th2 pathway where IgE acts upon cells bearing high-affinity IgE receptors to effect release of allergic mediators such as histamine, proteases, cytokines (TNF-α, IL-4, IL-13), and lipid mediators including prostaglandin D2 and leukotrienes B4 and E4.[29] Omalizumab is a humanized IgG1 kappa monoclonal antibody, with <5% murine complementarity-determining region, that binds to circulating free IgE with affinity comparable to the binding of IgE to its high-affinity receptor. The binding site for omalizumab on IgE is the same third constant heavy chain domain epitope that is the site for IgE binding to its receptors; thus omalizumab does not bind to IgE that is already bound to either high- or low-affinity receptors on inflammatory cells, as the ligand epitope are hidden. Competitive binding of omalizumab to IgE results in formation of IgE–anti-IgE complexes, primarily 2:2 tetramers and 3:3 hexamers, that do not activate complement and are slowly cleared by the reticuloendothelial system. Omalizumab at a concentration of 2–100 times basal IgE level results in >99% of IgE being complexed, leaving <1% available for binding to IgE receptors, thereby ablating a trigger of the allergic reaction cascade. This process also results in eventual downregulation of receptors and IgE production.[30] By reducing early- and late-phase reactions to allergic stimuli, omalizumab was posited to have the potential to prevent allergen-induced asthma exacerbations.[31] To reduce free IgE sufficiently, a dosing nomogram of omalizumab based on basal IgE level and weight was devised to ensure a minimum dose of 0.016 mg/kg/IgE/month.[32] Omalizumab is administered subcutaneously every 2–4 weeks.[33]

The early pivotal randomized double-blind placebo-controlled trials with omalizumab focused on adolescents and adults with established moderate or severe asthma and allergic sensitization (positive immediate skin test) to at least one perennial allergen (dust mite, cockroach, and/or cat or dog dander) requiring moderate or high doses of inhaled corticosteroids with or without long-acting bronchodilators.[34–40] Thus, fungal-associated asthma was not specifically assessed in these trials. In a meta-analysis of seven such trials lasting 6–12 months each, with pooled evaluation of 2,511 omalizumab and 1,797 control subjects in which asthma exacerbation was the primary endpoint in six trials, omalizumab was shown to reduce exacerbations by 38.3%. Health utilization decreased concomitantly (e.g., emergency room visits by 60% and hospital admissions by 51%) and quality of life scores improved.[40,41] Two pediatric trials in children 6–12 years old ($n = 609$ omalizumab, 301 placebo) with similar selection criteria and treatment length showed safety and comparable endpoint improvements.[42–44]

Recently, a 60-week controlled trial in young inner city 6- to 20-year-olds with predominantly moderate to severe asthma confirmed these benefits and also demonstrated a marked reduction in seasonal exacerbations, suggesting omalizumab reduces exacerbations and symptoms caused by seasonal changes that might be related to pollen or mold exposures (although these were not measured) and/or interactive effects on viral triggers.[45,46] This raises the possibility of seasonal rather than ongoing omalizumab therapy as a potential subject of study, as the maximum effect was observed within one month. A second recent study focused on inadequately controlled severe asthma in adults despite optimal current NIH guideline pharmacotherapy (National Asthma Education and Prevention Program Expert Panel Report 3E, steps 5 and 6);[47] in this 48-week study ($n = 427$ omalizumab, 423 placebo) exacerbations were reduced by 25% and other endpoints such as symptoms, rescue medication use, and quality of life scores also improved.[48] Interestingly in this study a noninvasive measure of pulmonary inflammation, exhaled nitric oxide concentration, also decreased with omalizumab, confirming cellular anti-inflammatory effects.[31,48,49]

Overall, the adverse effect profile of omalizumab is quite good, with rare anaphylactic reactions noted in 0.1–0.2% of recipients, which has led to recommendations that dosing be done under medical supervision with two hours of observation postinjection. There are also ongoing concerns about rare risks of malignancy (0.5% in recipients vs. 0.2% in controls) or possible cardiovascular/cerebrovascular adverse events.[50] The major limiting factors in its use, besides the specificity of its target, are the inconvenience of physician-supervised injections and the issue of pharmacoeconomic justification, that is, health benefits outweighing the high cost; thus, patient selection is key, and should focus on those with severe allergic asthma. To date omalizumab has not been specifically studied in SAFS, which appears to respond to add-on azole antifungal therapy.[8] However, because of the toxicity of prolonged systemic glucocorticosteroid therapy and often inadequate control of ABPA with combination steroid-azole therapy, omalizumab is increasingly used in treatment of ABPA.[51] To date 64 omalizumab-treated ABPA patients have been reported in abstracts and peer reviewed publications, with reduced exacerbations and systemic steroid burden being the main benefits of therapy.[52–60] However, no placebo-controlled trials have been completed. Two recent open-label

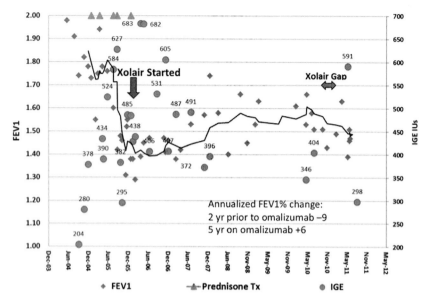

Figure 2. FEV1 and prednisone history. Long-term response to omalizumab in a patient with cystic fibrosis and allergic bronchopulmonary aspergillosis. This patient, currently 23 years old, had ABPA diagnosed in 2000 and was treated conventionally with prednisone and itraconazole, in addition to ongoing treatments for cystic fibrosis and asthma. After a period of accelerated lung function decline, despite several toxic courses of prednisone, she was started on omalizumab (Xolair®, Genentech). Subsequent lung function has been stabilized for six years, with a reduction in IgE and no further need for prednisone. An attempt to discontinue omalizumab in late 2010 resulted in an exacerbation, decline in lung function, and rise in IgE, which resolved with reinstitution of omalizumab. Green triangles = prednisone courses; orange circles = IgE (IU/mL) with values shown; blue diamonds = forced expiratory volume in one second (FEV1, L); solid line = rolling 6-month average FEV1.

series from Spain and France (pooled $n = 34$, 2 with CF-ABPA) showed significant reductions in exacerbations and oral steroid doses.[58,59] A major caveat for this approach, however, is the very high basal IgE levels in these patients, driving a need for concomitantly high doses of omalizumab—up to 600 mg weekly have been employed. Illustrative results of treatment of a patient with cystic fibrosis and ABPA are shown in Figure 2; in this patient accelerated decline in lung function and frequent exacerbations, despite steroid and azole therapy, were halted and stability was maintained, without adverse effects, by long-term omalizumab therapy.

Th2 cytokine inhibition

Although omalizumab is thus far the only biologic agent licensed for treatment of asthma, studies have shown that its real-world effectiveness is limited, with up to 40% of severe asthmatics being nonresponders—in terms of gaining asthma control.[61–63] However, the complexity of the Th2 pathway offers a rich variety of further potential targets for treatment, as well as potential limitations on highly specific agents (Fig. 1).[18,19,26,28,50] Only

those biologics whose evaluation reached the stage of mid-to-late (i.e., phase 2 or 3) clinical trials—where multidose safety and at least exploratory clinical endpoint efficacy measures were obtained—are discussed below. Several comprehensive reviews, including other agents and approaches, are available.[64–69]

The Th2 pathway includes an important component of eosinophilic recruitment and activation.[70] Interleukin 5 is known to play a central role in eosinophil differentiation, maturation, and survival. Early controlled clinical trials of monoclonal anti-IL-5 antibodies in patients with mild or moderate asthma demonstrated reductions in blood and sputum eosinophils but little clinical effect.[71] Subsequent trials have focused on selection of a severe adult-onset asthma with persistent sputum eosinophilia despite high-dose inhaled or systemic corticosteroid therapy phenotype, probably representing ~5% of adult asthmatics.[70] Haldar *et al.* studied 61 such patients (29 active, 32 placebo) who received monthly anti-IL-5 for a year and showed that treatment lowered exacerbation rate and increased quality of life score but did not affect pulmonary function.[72] Nair *et al.*, using the same antibody, studied 20 subjects (9 active, 11 placebo) in a 26-week protocol and also found that active treatment lowered exacerbation rate, improved symptoms and quality of life score, and allowed systemic steroid dose reduction.[73] Castro *et al.*, using another anti-IL-5 antibody, studied 106 patients (53 active, 53 placebo) for 16 weeks. In this shorter study, lung function and quality of life improved on active treatment, with a trend toward reduced exacerbations.[74] Thus, the current evidence suggests that there is a severe asthma phenotype that is responsive to anti-IL-5 therapy. Whether this group might include asthmatics with fungal sensitivity or ABPA remains to be seen.

A second major component of the Th2 pathway proximal to IgE induction is the action of IL-4 and IL-13 in furthering asthmatic pathology.[75] IL-4 and IL-13 have features both distinct and in common; both cytokines act in part via the IL-4 receptor alpha chain (IL-4Rα) of heterodimer receptor cell signaling ligands. Although IL-4 promotes differentiation and proliferation of CD4⁺ Th2 cells and production of IgE from B cells, IL-13 appears crucial in inducing and sustaining airway hyperreactivity, mucus secretion, and remodeling. Earlier studies using either monoclonal antibodies to IL-4 or a soluble IL-4 receptor were disappointing, although the reasons remain obscure.[67] More recent studies have focused on inhibiting IL-13, either via anti-IL-13 antibodies[76] or interruption of receptor-mediated signaling. A monoclonal antibody to IL-4Rα has shown some activity in patients with more severe asthma phenotype.[77] A similar approach using a mutated nonagonistic IL-4 molecule that competitively binds the IL-4Rα also shows promise.[19] Most impressively, a 24-week controlled trial by Corren *et al.* of an anti-IL-13 monoclonal antibody examined subjects ($n = 107$ active, 112 placebo) with asthma poorly controlled on inhaled corticosteroids.[78] Importantly, in this trial before randomization, subjects were stratified for the Th2 endotype by total IgE level (>100 IU/mL) and blood eosinophil count (>140/mL). Later in the study, serum periostin levels were added as an additional surrogate to examine the Th2-high and -low groups (periostin being an IL-13–induced epithelial product that appears to contribute to airway remodeling).[25] Anti-IL-13 improved pulmonary function; this effect was attributable to positive responses in the Th2-high

subgroup. This important study appears to validate a vital physiological role for IL-13 in the Th2 pathway in asthma, and thus offers an attractive and important target for further clinical trials.[75] Here, as with IL-5, the role of IL-13 inhibition in fungal asthma and APBA is currently undefined and merits investigation.

Unfortunately, amelioration of asthma by antibodies to TNF-α and IL-25Rα (CD25) has been outweighed by their toxic side effects, which precluded further development.[79,80]

Other strategies

Inhibition of Th2 cytokines is but one general approach to biological control of asthma. It is also possible to target cells directly. A promising approach is to use antisense oligodeoxynucleotides (ODNs) to target specific RNA sequences and downregulate transcription of specific proteins playing a role in asthma pathogenesis or pathophysiology. The eosinophil, as a major component of several asthma phenotypes, has been selected for study by development of ODNs against several proteins, including the CCR3 chemokine receptor, which has been correlated with asthma severity, and the common beta chain (CD131) of the heterodimeric receptors for GM-CSF, IL-3, and IL-5, all of which are eosinophil growth

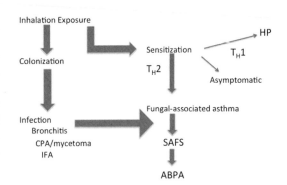

Figure 3. A hypothetical schematic representation of a Th2 pathway spectrum of clinical respiratory disease associated with fungi. Inhalation of fungal conidia or fragments can lead to allergic sensitization and simple fungal asthma. It is likely that in cases of poorly controlled asthma as well as cystic fibrosis, defects in airway host defense lead to germination of conidia and exposure of the host innate defenses to hyphal allergens and proteases that lead to more severe adaptive Th2 responses and granulocytic inflammation, manifesting clinically in severe asthma with fungal sensitization and allergic bronchopulmonary aspergillosis.

factors (in addition to having other activities).[81] An inhalable formulation of a combination small molecule containing two ODNs for CCR3 and CD131 has shown, in a controlled crossover four-day trial, the ability to inhibit sputum eosinophilia and allergen-induced asthmatic responses.[82,83] Further studies seem warranted to determine if these early observations translate into a safe and clinically effective approach.

Finally, it is vital to note that technologically advanced solutions may not necessarily provide the only, or even best, paths to biologic therapy. Fungal allergy and ABPA have been shown to be dependent upon respiratory epithelial cell activation and secretion of innate mediators (such as thymic stromal lymphopoeitin, IL-17, IL-25, and IL-33) that influence dendritic cells to secrete Th2-polarizing chemokines, such as CCL17 and CCL22.[18,84,85] Kreindler et al. demonstrated that dendritic cell orchestration of the Th2 pathway occurs via an OX40 ligand-dependent process that is downregulated by vitamin D.[86] Thus, vitamin D supplementation may prove beneficial in preventing or treating fungal allergy and ABPA, a possibility that is currently being tested in patients with CF and ABPA (ClinTrial.gov identifier NCT01222273).

In conclusion, asthma is heterogeneous. Fungal asthma, severe asthma with fungal sensitization, and ABPA are forms of an allergic or Th2 asthma endotype that manifest clinically along a phenotypic severity spectrum (Fig. 3). Omalizumab, or anti-IgE, is an effective biological approach to allergic asthma, including ABPA. Other biologicals targeting elements of the Th2 pathway, such as IL-13, IL-5, and eosinophils, show promise for selected severe Th2 endotype fungal asthma and ABPA patients. Controlled trials of biologics are needed in fungal asthma, severe asthma with fungal sensitization and allergic bronchopulmonary aspergillosis.

Conflicts of interest

The author declares no conflicts of interest.

References

1. Cookson, W. 1999. The alliance of genes and environment in asthma and allergy. *Nature* **402**(Suppl.): B5–B11.
2. Knutsen, A.P., R.K. Bush, J.G. Demain, *et al.* 2012. Fungi and allergic lower respiratory tract disease. *J. Allergy Clin. Immunol.* **129**: 280–291.
3. Zureik, M., C. Nuekirch, B. Leynaert, *et al.* 2002. Sensitization to airborne moulds and severity of asthma: cross sectional study from European community respiratory health survey. *Br. Med. J.* **325**: 411–414.
4. Denning, D.W., B.R. O'Driscoll, C.M. Hogaboam, *et al.* 2006. The link between fungi and asthma—a summary of the evidence. *Eur. Respir. J.* **27**: 615–626.
5. Agarwal, R. & D. Gupta. 2011. Severe asthma and fungi: current evidence. *Med. Mycol.* **49**(Suppl. 1): S150–S157.
6. Agarwal, R., A.N. Aggarwarl, D. Gupta & S.K. Jindal. 2009. Aspergillus hypersensitivity and allergic bronchopulmonary aspergillosis in patients with bronchial asthma: systematic review and meta-analysis. *Int. J. Tuberc. Lung Dis.* **13**: 936–944.
7. Fairs, A., J. Agbetile, B. Hargadon, *et al.* 2010. IgE sensitization to Aspergillus fumigatus is associated with reduced lung function in asthma. *Am. J. Respir. Crit. Care Med.* **182**: 1362–1368.
8. Denning, D.W., B.R. O'Driscoll, G. Powell, *et al.* 2009. Randomized controlled trial of oral antifungal treatment for severe asthma with fungal sensitization. The fungal asthma sensitization (FAST) study. *Am. J. Respir. Crit. Care Med.* **179**: 11–18.
9. Moss, R.B. 2009. Allergic bronchopulmonary aspergillosis. In *Aspergillus Fumigatus and Aspergillosis.* J.P. Latge & W.J. Steinbach, Eds.: 333–350. American Society of Microbiology Press. Washington, DC.
10. Hogan, C. & D.W. Denning. 2011. Allergic bronchopulmonary aspergillosis and related allergic sydromes. *Semin. Respir. Crit. Care Med.* **32**: 682–692.
11. Robinson, D.S., Q. Hamid, S. Ying, *et al.* 1992. Predominant TH2-like bronchoalveolar T-lymphocyte population in atopic asthma. *N. Eng. J. Med.* **326**: 298–304.
12. Salvi, S.S., S. Babu & S.T. Holgate. 2001. Is asthma really due to a polarized T cell response toward a helper T cell type 2 phenotype? *Am. J. Respir. Crit. Care Med.* **164**: 1343–1346.
13. Haldar, P., I.D. Pavord, D.E. Shaw, *et al.* 2008. Cluster analysis and clinical asthma phenotypes. *Am. J. Respir. Crit. Care Med.* **178**: 218–224.
14. Moore, W.C., D.A. Meyers, S.E. Wenzel, *et al.* 2010. Identification of asthma phenotypes using cluster analysis in the severe asthma research program. *Am. J. Respir. Crit. Care Med.* **181**: 315–323.
15. Fitzpatrick, A.M., W.G Teague, D.A. Meyers, *et al.* 2011. Heterogeneity of severe asthma in childhood: confirmation by cluster analysis of children in the NIH/NHLBI severe asthma research program. *J. Allergy Clin. Immunol.* **127**: 382–389.
16. Wenzel, S. 2006. Asthma: defining of the persistent adult phenotypes. *Lancet* **368**: 804–813.
17. Lötvall, J., C.A. Adkis, L.B. Bacharier, *et al.* 2011. Asthma endotypes: a new approach to classification of disease entities within the asthma syndrome. *J. Allergy Clin. Immunol.* **127**: 355–360.
18. Kim, H.Y., R.H DeKruyff & D.T. Umetsu. 2010. The many paths to asthma: phenotype shaped by innate and adaptive immunity. *Nature Immunol.* **11**: 577–584.
19. Wenzel, S. 2012. Severe asthma: from characteristics to phenotypes to endotypes. *Clin. Exp. Allergy* **42**: 650–658.
20. Editorial. 2006. A plea to abandon asthma as a disease concept. *Lancet* **368**: 705.

21. Spycher, B.D., M. Silverman & C.E. Kuehni. 2010. Phenotypes of childhood asthma: are they real? *Clin. Exp. Allergy* **40:** 1130–1141.

22. McGrath, K.W., N. Icitovic, H.A. Boushey, *et al.* 2012. A large subgroup of mild-to-moderate asthma is persistently noneosinophilic. *Am. J. Crit. Care Med.* **185:** 612–619.

23. Kennedy, J.L., P.W. Heymann & T.A.E. Platts-Mills. 2012. The role of allergy in severe asthma. *Clin. Exp. Allergy* **42:** 659–669.

24. Anderson, G.P. 2008. Endotyping asthma: new insights into key pathogenic mechanisms in a complex, heterogeneous disease. *Lancet* **372:** 1107–1119.

25. Woodruff, P.G, B. Modrek, D.F. Choy, *et al.* 2009. T-helper type 2-driven inflammation defines major subphenotypes of asthma. *Am. J. Respir. Crit. Care Med.* **180:** 388–395.

26. Bhakta, N.R. & P.G. Woodruff. 2011. Human asthma phenotypes: from the clinic, to cytokines, and back again. *Immunol. Rev.* **242:** 220–232.

27. Gernez, Y., C.E. Dunn, C. Everson, *et al.* 2012. Blood basophils from cystic fibrosis patients with allergic bronchopulmonary aspergillosis are primed and hyper-responsive to stimulation by *Aspergillus* allergens. *J. Cystic Fibrosis* http://dx.doi.org/10.1016/j.jcf.2012.04.008.

28. Bosnjak, B., B. Stelzmueller, K.J. Erb & M.M. Epstein. 2011. Treatment of allergic asthma: modulation of Th2 cells and their responses. *Respir. Res.* **12:** 144.

29. Holgate, S., T. Casale, S. Wenzel, *et al.* 2005. The anti-inflammatory effects of omalizumab confirm the central role of IgE in allergic inflammation. *J. Allergy Clin. Immunol.* **115:** 459–465.

30. Chang, T.W. 2000. The pharmacological basis of anti-IgE therapy. *Nature Biotech.* **18:** 157–162.

31. Holgate, S.T., R. Djukaovic, T. Casale & J. Bousquet. 2005. Anti-immunoglobulin E treatment with omalizumab in allergic diseases: an update on anti-inflammatory activity and clinical efficacy. *Clin. Exp. Allergy* **35:** 408–416.

32. Hochhaus, G., L. Brookman, H. Fox, *et al.* 2003. Pharmacodynamics of omalizumab: implications for optimised dosing strategies and clinical efficacy in the treatment of allergic asthma. *Curr. Med. Res. Opin.* **19:** 491–498.

33. Strunk, R.C. & G.R Bloomberg. 2006. Omalizumab for asthma. *N. Engl. J. Med.* **354:** 2689–2695.

34. Busse, W., J. Corren, B.Q. Lanier, *et al.* 2001. Omalizumab, anti-IgE recombinant humanized monoclonal antibody, for the treatment of severe allergic asthma. *J. Allergy Clin. Immunol.* **108:** 184–190.

35. Solèr, M., J. Matz, R. Townley, *et al.* 2001. The anti-IgE antibody omalizumab reduces exacerbations and steroid requirement in allergic asthmatics. *Eur. Respir. J.* **18:** 254–261. [Erratum in: *Eur. Respir. J.* **18:** 739–740].

36. Ayres, J.G., B. Higgins, E.R. Chilvers, *et al.* 2004. Efficacy and tolerability of anti-immunoglobulin E therapy with omalizumab in patients with poorly controlled (moderate-to-severe) allergic asthma. *Allergy* **59:** 701–708.

37. Vignola, A.M., M. Humbert, J. Bousquet, *et al.* 2004. Efficacy and tolerability of anti-immunoglobulin E therapy with omalizumab in patients with concomitant allergic asthma and persistent allergic rhinitis: SOLAR. *Allergy* **59:** 709–717.

38. Holgate, S.T., A.G. Chuchalin, J. Hébert, *et al.* 2004. Efficacy and safety of a recombinant anti-immunoglobulin E antibody (omalizumab) in severe allergic asthma. *Clin. Exp. Allergy* **34:** 632–638.

39. Humbert, M., R. Beasley, J. Ayres, *et al.* 2005. Benefits of omalizumab as add-on therapy in patients with severe persistent asthma who are inadequately controlled despite best available therapy (GINA 2002 step 4 treatment): INNOVATE. *Allergy* **60:** 309–316.

40. Bousquet, J., P. Cabrera, N. Berkman, *et al.* 2005. The effect of treatment with omalizumab, an anti-IgE antibody, on asthma exacerbations and emergency medical visits in patients with severe persistent asthma. *Allergy* **60:** 302–308.

41. Finn, A., G. Gross, J. van Bavel, *et al.* 2003. Omalizumab improves asthma-related quality of life in patients with severe asthma. *J. Allergy Clin. Immunol.* **111:** 278–284.

42. Milgrom, H., W. Berger, A. Nayak, *et al.* 2001. Treatment of childhood asthma with anti-immunoglobulin E antibody (omalizumab). *Pediatrics* **108:** e36.

43. Berger, W., N. Gupta, M. McAlary & A. Fowler-Taylor. 2003. Evaluation of long-term safety of the anti-IgE antibody, omalizumab, in children with allergic asthma. *Ann. Allergy Asthma Immunol.* **91:** 182–188.

44. Lanier, B., T. Bridges, M. Kulus, *et al.* 2009. Omalizumab for the treatment of exacerbations in children with inadequately controlled allergic (IgE-mediated) asthma. *J. Allergy Clin. Immunol.* **124:** 1210–1216.

45. Busse, W.W., W.J. Morgan, P.J. Gergen, *et al.* 2011. Randomized trial of omalizumab (anti-IgE) for asthma in inner-city children. *N. Eng. J. Med.* **364:** 1005–1015.

46. Busse, W.W., R.F. Lemanske, Jr. & J.E. Gern. 2010. Role of viral respiratory infections in asthma and asthma exacerbations. *Lancet* **376:** 826–834.

47. National Asthma Education and Prevention Program, National Heart, Lung, and Blood Institute, National Institutes of Health. 2007. *Expert Panel Report 3: Guidelines for the Diagnosis and Management of Asthma*; Full Report. NIH publication. Bethesda, MD. US Dept of Health and Human Services: August 2007. 07–4051.

48. Hanania, N.A., O. Alpan, D.L Hamilos, *et al.* 2011. Omalizumab in severe allergic asthma inadequately controlled with standard therapy. A randomized trial. *Ann. Intern. Med.* **154:** 573–582.

49. Djukanovic, R., S.J. Wilson, M. Kraft, *et al.* 2004. Effects of treatment with anti-immunoglobulin E antibody omalizumab on airway inflammation in allergic asthma. *Am. J. Respir. Crit. Care Med.* **170:** 583–593.

50. Levine, S.J. & S.E. Wenzel. 2010. The role of Th2 immune pathway modulation in the treatment of severe asthma and its phenotypes: are we getting closer? *Ann. Int. Med.* **152:** 232–237.

51. Moss, R.B. 2008. Management of allergic aspergillosis. *Curr. Fungal Infect. Rep.* **2:** 87–93.

52. Van der Ent, C.K., H. Hoekstra & G.T. Rijkers. 2007. Successful treatment of allergic bronchopulmonary aspergillosis with recombinant anti-IgE antibody. *Thorax* **62:** 276–277.

53. Zirbes, J.M. & C.E. Milla. 2008. Steroid-sparing effect of omalizumab for allergic bronchopulmonary aspergillosis and cystic fibrosis. *Pediatr. Pulmonol.* **43:** 607–610.

54. Kanu, A & K. Patel. 2008. Treatment of allergic bronchopulmonary aspergillosis (ABPA) in CF with anti-IgE antibody (omalizumab). *Pediatr. Pulmonol.* **43:** 1249–1251.

55. Lebecque, P. & C. Pilette. 2009. Omalizumab for treatment of ABPA excerbations in CF patients. *Pediatr. Pulmonol.* **44:** 516.

56. Brinkmann, F., N. Schwerk, G. Hansen & M. Ballman. 2009. Steroid dependency despite omalizumab treatemtn of ABPA in cystic fibrosis. *Allergy* **65:** 134–135.

57. Lin, R.Y., S. Sethi & G.A. Bhargave. 2010. Measured immunoglobulin E in allergic bronchopulmonary aspergillosis treated with omalizumab. *J. Asthma* **47:** 942–945.

58. Pérez-de-Llano, L.A., M.C. Vennera, A. Parra, *et al.* 2011. Effects of omalizumab in Aspergillus-associated airway disease. *Thorax* **66:** 539–540.

59. Tillie-Leblond, I., P. Germaud, C. Leroyer, *et al.* 2011. Allergic bronchopulmonary aspergillosis and omalizumab. *Allergy* **66:** 1254–1256.

60. Genentech. Xolair—use in the treatment of allergic bronchopulmonary aspergillosis. US-XOL-S040, V.13.0. 16 Aug 2011.

61. Niven, R., K.F. Chung, Z. Panahlo, *et al.* 2008. Effectiveness of omalizumab in patients with inadequately controlled severe persistent allergic asthma: an open-label study. *Respir. Med.* **102:** 1371–1378.

62. Humbert, M., W. Berger, G. Rapatz & F. Turk. 2008. Add-on omalizumab improves day-to-day symptoms in inadequately controlled severe persistent allergic asthma. *Allergy* **63:** 592–596.

63. Brusselle, G., A. Michils, R. Louis, *et al.* 2009. "Real-life" effectiveness of omalizumab in patient with severe persistent allergic asthma: the PERSIST study. *Respir. Med.* **103:** 1633–1642.

64. O'Byrne, P.M. 2006. Cytokines or their antagonists for the treatment of asthma. *Chest* **130:** 244–250.

65. Adcock, I.M, G. Caramori & K.F. Chung. 2008. New targets for drug development in asthma. *Lancet* **372:** 1073–1087.

66. Thomson, N.C., R. Chaudhuri & M. Spears. 2011. Emerging therapies for severe asthma. *BMC Medicine* **9:** 102.

67. Corren, J. 2011. Cytokine inhibition in severe asthma: current knowledge and future directions. *Curr. Opinion Pulm. Med.* **17:** 29–33.

68. Hansbro, P.M., G.E. Kaiko & P.S. Foster. 2011. Cytokine/anti-cytokine therapy—novel treatments for asthma? *Br. J. Pharmacol.* **163:** 81–95.

69. O'Byrne, P.M., N. Naji & G.M. Gauvreau. 2012. Severe asthma: future treatments. *Clin. Exp. Allergy* **42:** 706–711.

70. Wenzel, S.W. 2009. Eosinophils in asthma—closing the loop or opening the door? *N. Engl. J. Med.* **360:** 1026–1029.

71. Corren, J. 2011. Anti-intereukin-5 antibody therapy in asthma and allergies. *Curr. Opin. Allergy Clin. Immunol.* **11:** 565–570.

72. Haldar, P., C.E. Brightling, B. Hargadon, *et al.* 2009. Mepolizumab and exacerbations of refractory eosinophilic asthma. *N. Eng. J. Med.* **360:** 973–984.

73. Nair, P., M.M. Pizzichini, M. Kjarsgaard, *et al.* 2009. Mepolizumab for prednisone-dependent asthma with sputum eosinophilia. *N. Eng. J. Med.* **360:** 985–993.

74. Castro, M., S. Mathur, F. Hargreave, *et al.* 2011. Reslizumab for poorly controlled, eosinophilic asthma: a randomized, placebo-controlled study. *Am. J. Respir. Crit. Care Med.* **184:** 1125–1132.

75. Kraft, M. 2011. Asthma phenotypes and interleukin-13—moving closer to personalized medicine. *N. Eng. J. Med.* **365:** 1141–1144.

76. Gauvreau, G.M., L.P. Boulet, D.W. Cockroft, *et al.* 2011. Effects of interleukin-13 blockade on allergen-induced airway responses in mild atopic asthma. *Am. J. Respir. Crit. Care Med.* **183:** 1007–1014.

77. Corren, J., W. Busse, E.O. Meltzer, *et al.* 2010. A randomized, controlled, phase 2 study of AMG 317, an IL-4Ralpha antagonist, in patients with asthma. *Am. J. Respir. Crit. Care Med.* **181:** 788–796.

78. Corren, J., R.F. Lemanske, Jr., N. A. Hanania, *et al.* 2011. Lebrikizumab treatment in adults with asthma. *N. Eng. J. Med.* **365:** 1088–1098.

79. Busse, W.W., E. Israel, H.S. Nelson, *et al.* 2008. Daclizumab improves asthma control in patients with moderate to severe persistent asthma: a randomized, controlled trial. *Am. J. Respir. Crit. Care Med.* **178:** 1002–1008.

80. Wenzel, S.E., P.J. Barnes, E.R. Bleecker, *et al.* 2009. A randomized, double-blind, placebo-controlled study of tumor necrosis factor-alpha blockade in severe persistent asthma. *Am. J. Respir. Crit. Care Med.* **179:** 549–558.

81. Wegmann, M. 2011. Targeting eosinophil biology in asthma therapy. *Am. J. Respir. Cell. Mol. Biol.* **45:** 667–674.

82. Gauvreau, G.M., L.P. Boulet, D.W. Cockroft, *et al.* 2008. Antisense therapy against CCR3 and the common beta chain attenuates allergen-induced eosinophilic responses. *Am. J. Resp. Crit. Care Med.* **177:** 952–958.

83. Imaoka, H., H. Campbell, I. Babirad, *et al.* 2011. TPI ASM8 reduces eosinophil progenitors in sputum after allergen challenge. *Clin. Exp. Allergy* **41:** 1740–1746.

84. Rapaka, R.R & J.K. Kolls. 2009. Pathogenesis of allergic bronchopulmonary aspergillosis in cystic fibrosis: current understanding and future directions. *Med. Mycol.* **47(Suppl.1):** S331–S337.

85. Fahy, J.V. & R.M. Locksley. 2011. The airway epithelium as a regulator of Th2 responses in asthma. *Am. J. Crit. Care Med.* **184:** 390–392.

86. Kreindler, J.L., C. Steele, N. Nguyen, *et al.* 2010. Vitamin D3 attenuates Th2 reponses to Aspergillus fumigatus mounted by CD4+ T cells from cystic fibrosis patients with allergic bronchopulmonary aspergillosis. *J. Clin. Invest.* **120:** 3242–3254.

Ann. N.Y. Acad. Sci. ISSN 0077-8923

ANNALS OF THE NEW YORK ACADEMY OF SCIENCES

Issue: *Advances Against Aspergillosis*

Multifocal pulmonary aspergillomas: case series and review

Matthew Pendleton[1] and David W. Denning[2]

[1]Ninewells Hospital and Medical School, University of Dundee, Dundee, United Kingdom. [2]The National Aspergillosis Centre, University Hospital of South Manchester, The University of Manchester, Manchester Academic Health Science Centre, Manchester, United Kingdom

Address for correspondence: David W. Denning, Education & Research Centre, University Hospital of South Manchester, Southmoor Road, Manchester, M23 9LT, United Kingdom. ddenning@manchester.ac.uk

Multifocal lung parenchymal cavities containing multiple aspergillomas are not well-recognized features of chronic pulmonary aspergillosis (CPA). We identified five patients with multiple cavities containing fungal balls from our accumulated cohort of ~300 patients with CPA. Corticosteroid exposure and radiological and serological characteristics were assessed. The patients, aged 19–55 years, developed 3–11 cavities (or nodules in one case) with thin walls, usually within the lung parenchyma, with very limited pleural involvement. Four had asthma (severe in three) and one had cystic fibrosis; three had allergic bronchopulmonary aspergillosis. All patients had received corticosteroids orally or by inhalation. Four patients had elevated *Aspergillus* IgG antibodies; one had elevated *Aspergillus*-specific IgE. Three patients developed azole resistance on antifungal therapy, after benefit, one of whom underwent a successful bilateral lung transplant, later complicated by a fatal mycotic cerebral aneurysm. Multiple aspergillomas is a new distinct manifestation of CPA. The lack of inflammatory response and the distribution of the cavities in the lungs are remarkable. *Aspergillus* nodules could evolve into cavities with aspergillomas. The management and development of azole resistance in these patients is problematic.

Keywords: fungal infection; chronic pulmonary aspergillosis; *Aspergillus fumigatus*; fungal ball; chronic necrotizing pulmonary aspergillosis

Introduction

Daily exposure to airborne *Aspergillus* is an inescapable fact of life. Human lungs are usually not sterile with respect to fungi.[1–3] It is therefore not surprising that either structural pulmonary or immunological deficiency allows this rapidly growing filamentous fungus to infect localized areas of lung. Indeed patients with allergic bronchopulmonary aspergillosis (ABPA)[4,5] and those with severe asthma are usually colonized in the airways by *Aspergillus fumigatus*,[6] as are many patients with asthma.[6]

In 1842, John Hughes Bennett, a medical microscopist of great renown, found what he thought was a vegetable (resembling *Aspergillus*) in the sputum of a patient.[7] He noted; "the left lung was found studded with cavities of different sizes, some of which communicated, by fistulous openings, with the cavity of the pleura. Several of the smaller cavities were partly filled with soft tuberculous matter, readily separable from the lining membrane." This was the first description of chronic pulmonary aspergillosis (CPA), indeed the first description of any form of pulmonary aspergillosis. CPA is generally a slowly progressive destructive lung disease characterized by pulmonary and/or systemic symptoms, cavitation of the lung, and positive *Aspergillus*-specific IgG antibodies detectable in serum.[8–10] Existing cavities increase the likelihood of *Aspergillus* infection, and preceding diseases include tuberculosis and sarcoidosis, among others.[11] We, and others years before, have also observed new cavity formation during the course of untreated chronic pulmonary aspergillosis; thus, a preexisting cavity is not a sine qua non of this disease. Cavities may or may not contain an aspergilloma.

The first formal description of an aspergilloma was in 1938 by Deve, who referred to it as a "mega-mycetoma."[12] An aspergilloma is a usually spherical mass of tangled hyphae often with a

doi: 10.1111/j.1749-6632.2012.06827.x

Ann. N.Y. Acad. Sci. 1272 (2012) 58–67 © 2012 New York Academy of Sciences.

necrotic core, visualized radiologically;[13] the vast majority are caused by *A. fumigatus*. Radiological features preceding the development of an aspergilloma include irregular cavity walls or loose material within a cavity that does not resemble a fungal ball. Most patients with aspergillomas have disease localized to one upper lobe of the lung or, rarely, both upper lobes. In this report, we describe multifocal nodules and cavities as an unusual manifestation of chronic pulmonary aspergillosis, many of the cavities containing aspergillomas. We attempt to infer the radiological progression of these manifestations of aspergillosis from patients' images and clinical course.

Patients and method

Patients were identified from the database of >500 individuals referred to the National Aspergillosis Centre (in the UK). We identified five patients with at least three multiple cavities and/or nodules in the parenchyma of the lung (Table 1), which were unusual in appearance compared with the normal appearance of chronic pulmonary aspergillosis and simple aspergilloma.

Medline and *Aspergillus*-related website searches in January 2011 using terms *multiple* and *aspergilloma* and *nodule* identified only reports of immunocompromised patients. A detailed trawl through >100 papers going back to 1938 and describing aspergilloma and chronic pulmonary aspergillosis identified six papers describing multiple aspergillomas.

Laboratory methods

Aspergillus IgG and IgE and total IgE were tested by ImmunoCap® (Phadia, Sweden), with cut-offs of 40 mg/L, 0.4 KIU/mL, and 100 KIU/mL. *Aspergillus* precipitin titers were measured by an in-house precipitins IgG assay.[14] Sputum was cultured for fungi using standard UK methods; DNA extraction from sputum and real-time PCR was performed as described previously.[15] Minimum inhibitory concentrations (MICs) to triazoles were determined by EUCAST methodology with a lower final inoculum concentration (0.5×10^5 as opposed to $1–2.5 \times 10^5$ cfu/mL) as the only methodological variation.[16] Antifungal therapeutic drug monitoring for all triazole therapy was as described elsewhere.[17]

Case histories

Patient 1

A 34-year-old female had poorly controlled asthma in 2008. Before her asthma deterioration, she was fit, attending the gym, and long distance swimming. She had had an accident at age 15 resulting in spinal cord transection and as a result is wheelchair bound. Following the West Country floods in January 2008, there was much external mold contamination, although none in the patient's house. In 2008, she was on oral antibiotics much of the year and received three courses of oral prednisolone (40 mg daily) for a week, with partial improvement, in addition to high dose inhaled steroids. In November 2008, a chest X-ray (CXR) was abnormal, the first radiological sign of CPA. Abnormal CXRs showed two possible balls in a bronchiectatic cavity in her left upper lobe (LUL), with a further smaller ball in her right upper lobe (RUL). On referral her total IgE was 300 mg/L, *A. fumigatus* IgE 3.3 KIU/L, but her *Aspergillus* IgG was >200 mg/L. A CT of her thorax (Fig. 1) confirmed multiple cavities containing "balls," and a pulmonary embolus for which she was started on warfarin, which was changed to dalteparin after an episode of hemoptysis. She was also started on voriconazole 200 mg twice daily (bd), which was well tolerated after the initial symptoms of visual disturbance and nausea, with satisfactory trough levels (4.7, 5.24 mg/L). Her *Aspergillus* IgG fell to 144 over the next year and she remained well, competing in long-distance races. In May 2011, she had a flare of her ABPA that required another course of corticosteroids and ended up being treated with nebulized Pulmicort® 1 mg bd and prednisolone 10 mg daily. An embolization procedure was then required for hemoptysis, suggesting therapeutic failure. Her *Aspergillus* IgG climbed to 529 mg/L and a strong *Aspergillus* PCR signal was detected in sputum, despite therapeutic voriconazole levels, confirming therapeutic failure and presumptive voriconazole resistance. Culture was negative.

Patient 2

This 19-year-old female was referred in 2009 after her sputum tested culture positive for *A. fumigatus* despite taking itraconazole. Her primary underlying diagnosis was cystic fibrosis (CF) complicated by diabetes mellitus and pancreatic insufficiency. She had been colonized with *A. fumigatus* from early teenage years and had serological

Table 1. Clinical details of the patients

No	Age sex	Underlying conditions	Presentation	Lobes affected	Asp IgG (max)	IgE (max)	Asp IgE (max)	Sputum culture	Drug Tx	Outcome
1	34/F	Asthma, ABPA, bronchiectasis	Abnormal CXR during an chronic asthma exacerbation	LUL, RUL, RLL	>200	1500	13.2	*A. fumigatus*	Initially itra, then vori 200 mg bd, (levels 4.7 and 5.24)	Clinically stable, but new hemoptysis and strong *Aspergillus* PCR signal on therapy, culture negative, resistance
2	19/F	CF, ABPA, diabetes pancreatic insufficiency	Persistently positive sputum culture positive and abnormal CT	LUL, RUL	161	110	3.7	Triazole-resistant *A. fumigatus*	Resistant to all azoles, anaphylaxis with Ambisome, maintained on voriconazole prior to lung transplant	Lung transplant May 2011, with intra-operative pleural torulidine, i.v. micafungin and inhaled AmB
3	47/F	Severe asthma requiring intensive care in 90s with fungal sensitization	R pneumotho-rax, cavitating lesions in both lungs	RLL, RUL, LUL	15	79	2.6	*A. fumigatus*	Vori 300 mg bd then 150 mg bd then 250 bd based on TDM	Stable, asthma controlled with fluticasone 500 bd.
4	55/M	Severe asthma, osteoporosis	Progressive SOB (three years), brown mucus plugs	RUL, LUL	78	639	18.3	Negative	Vori 200 mg bd (intolerant), 3 weeks course i.v. AmBisome® then itra (300 mg)	Chest stable
5	49/F	Moderate asthma, smoker	Right sided pleuritic chest pain, with consolidation that did not clear on antibiotics	RUL, LUL, RLL	165	2500	45.3	*A. fumigatus*	Pred 20 mg for 5 months with improvement, but deteriorated drastically when stopped; confirmed by CT (five new cavitating lesions on CT)	Asthma controlled on fluticasone and itra 200 mg bd (good levels)

ABPA, allergic bronchopulmonary aspergillosis; CPA, chronic pulmonary aspergillosis; CXR, chest X-ray; LUL, left upper lobe; RUL, right upper lobe; LLL, lower left lobe; RLL, right lower lobe; CT, computer tomography; CF, cystic fibrosis; ARDS, acute respiratory distress syndrome; PET, positron emission tomography; ANCA, antineutrophil cytoplasmic antibodies.

markers consistent with ABPA, which fluctuated. She was treated with a course of prednisolone, inhaled corticosteroids, occasional courses of prednisolone (once or twice annually), and itraconazole for many years. Due to her severe CF she had extensive bronchial dilatation particularly in the upper lobes, with many dilated bronchi and cavities containing aspergillomas (Fig. 2). A fungal

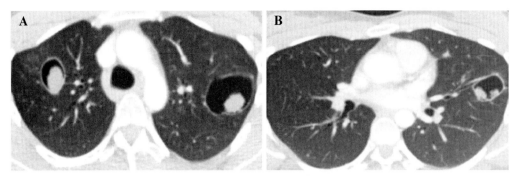

Figure 1. Patient 1. CT scan showing mild bronchiectasis with two thin-walled cavities bilaterally. In the right lung A is a section of the upper lobe, with B in the lower lobe showing the same cavity. Note that there are three separate aspergilloma in this cavity alone. Fungal balls measured 19 mm (left upper lobe, LUL), 28 mm (left upper lobe, LUL), 20, 18, and 9 mm (in same right upper lobe (RUL) cavity) and two others 20 mm and 23 mm.

culture (December 2009) grew two isolates of *A. fumigatus* that showed triazole resistance (itraconazole MIC >8 mg/L, voriconazole MIC >8 mg/L, posaconazole MIC 1 mg/L; and itraconazole MIC >8 mg/L, voriconazole MIC 2 mg/L, posaconazole MIC 0.5 mg/L). During a prior admission she had suffered urticarial rash and hypotension with AmBisome®, and so was skin tested and then challenged with conventional amphotericin B. She was successfully transplanted in May 2011, with a heart-lung bypass and did well initially. Her lungs were resected successfully without difficulty because there were few adhesions caused by the aspergillomas, as they were mostly intrapulmonary or bronchial, with little pleural involvement. Histopathology confirmed the presence of multiple fungal balls bilaterally. She was treated with daily micafungin and aerosolized amphotericin B postoperatively but grew two strains of *A. fumigatus*, one fully susceptible and one triazole resistant. Her first biopsy after transplantation showed moderately severe rejection that was treated with pulse steroids, and after a second biopsy (which did not show rejection) she developed severe ARDS and spent a month ventilated. She recovered from this and was discharged to a hospital near home. Unfortunately she suffered an intracranial hemorrhage two months posttransplantation related to a presumed mycotic aneurysm from which she did not recover. A coroner's autopsy was performed.

Patient 3
A 57-year-old female with chronic asthma that had been unusually difficult to control had been followed since an ICU admission involving allergy to *Aspergillus* in the 1990s. She had been markedly

steroid dependent, requiring multiple courses of prednisolone as well as 2 g fluticasone via a dry powder inhaler for asthma control. The long duration and symptomatic history are suggestive of long standing ABPA or severe asthma with fungal sensitization (SAFS) that had progressed to CPA. In August 2010 she suffered a right pneumothorax with cavitating lesions found bilaterally; this was confirmed with a CT of the thorax (Fig. 3). At referral, she was *Aspergillus* precipitins positive, her *Aspergillus* IgE was 13 kIU/L, and her total IgE was 890 kIU/L. She is currently on Seretide® 500 1 puff and voriconazole 250 mg both twice daily, which is well tolerated, with much better control of her asthma. There has been no significant change in her imaging over 12 months.

Patient 4
Patient 4 is a 55-year-old male who presented with progressive dyspnoea over three years, with an occasional productive cough of brown mucus "plugs," suggestive of ABPA. He had a 25-year history of steroid-dependant severe asthma, controlled by prednisolone 3.5 mg daily, that was complicated by osteoporosis and spondylosis (he remains on this dose despite efforts to decrease the minimum effective dosage). An investigation revealed an abnormal CXR; a CT showed multiple nodules bilaterally in the upper lobes, some cavitating (see Fig. 4). A CT-guided biopsy showed fungal hyphae and no evidence of malignancy; a subsequent lavage cultured *A. fumigatus*. His serology showed maximum levels of total IgE of 639 mg/L, *A. fumigatus* IgE of 18.3 KIU/L, and *Aspergillus* IgG of 78 mg/L. Testing for gamma interferon production in response

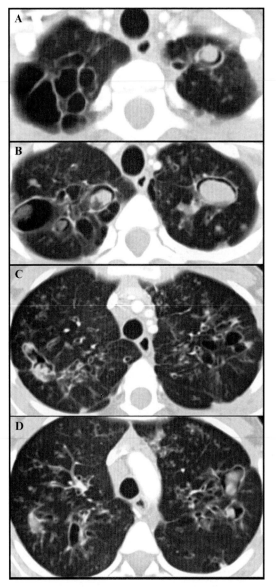

Figure 2. Patient 2. The figures above are images taken from a single CT scan at varying points in the upper lobes of this CF patient. (A) and (B) are two images in upper lobes of the lung, showing multiple thin-walled cavitations. Classic aspergilloma of clear halo and with no cavitation presents within the cavities bilaterally. (C) and (D) show widespread bronchiecstasis and cavitation. In the right lung is a single cavity displaying two fungal balls, however this may also occur in a vertical plane. Eleven fungal balls were visualized by imaging.

to multiple stimuli showed greatly reduced production but a normal response pathway, and he has since been started on gamma-IFN supplementation. Azole-resistant *A. fumigatus* (itraconazole MIC >8 mg/L, voriconazole MIC >8 mg/L, posaconazole

MIC 1 mg/L) was grown; he was given a three-week course of Ambisone for hemoptysis, which has now resolved. While there has been little change in his serology over two years, there has been increasing cavitation in his middle zone lesions, but none in his right upper lobe nodular lesions.

Patient 5

A 49-year-old female smoker initially presented with right-sided pleuritic chest pain (2006) and consolidation that did not clear on antibiotics. She was diagnosed with asthma in 1999, which was initially difficult to control but then settled; she currently takes fluticasone + salmeterol regularly. She worked next to a demolition site with, presumably, a higher risk of *Aspergillus* exposure. On referral she had at least three nodules, the largest being 3.5 cm in her right upper lobe. Several scans showed multiple aspergillomas in both upper lobes and right lower lobe. A PET scan showed low uptake. Her serology at referral was *Aspergillus* IgG 165 mg/L, total IgE 2500 kIU/L, and *A. fumigatus* IgE 45.3 kIU/L. Further follow-up revealed eosinophilia and a negative ANCA result. Last year she was treated with prednisolone 20 mg for five months due to radiological progression of aspergilloma to which she had a good response. However, soon after stopping treatment she had a dramatic deterioration in CT, showing five cavitating lesions (Fig. 5). Despite no prior history of hemoptysis recently some blood has been seen mixed with sputum, accompanied with left-sided pleuritic chest pain upon deterioration. Subsequent sputum culture was positive for *A. fumigatus.* She is currently being treated with itraconazole 200 mg BD, with satisfactory levels and well tolerated.

Results and discussion

The term *pulmonary aspergilloma* refers to the presence of solid mass of intertwined *Aspergillus* hyphae within an intrathoracic cavity, usually in the parenchyma of the lung, rarely in the pleura or bronchus.[11] Aspergillomas form when *Aspergillus* conidia adhere to the wall of a cavity, germinate, and gradually grow to form an amorphous mass.[18,19] Often, an affected pulmonary cavity wall has a cobblestone or shaggy appearance, sometimes with ulceration, prior to the overt formation of an aspergilloma.[19] Radiologically, an aspergilloma will have a distinct "halo" or crescent of air round part of the ball as it sits in a cavity. Sometimes

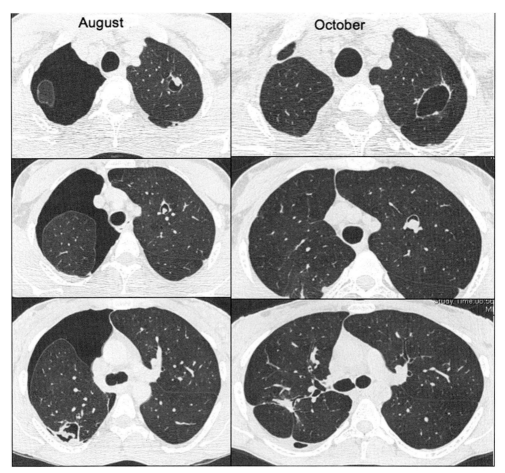

Figure 3. Patient 3. Images demonstrating pneumothorax, as a result of a right-sided cavity with a fungal ball perforating into the pleural cavity. Two other cavities with intraluminal material are present in the left upper lobe. Significant progression is seen in the three months to October.

aspergillomas are closely adherent to the cavity wall, but often they are mobile in the cavity.

What distinguishes the cases we describe above is the involvement of multiple lobes bilaterally with small to medium sized thin-walled cavities, typically without the usual pleural thickening and peripheral location. There was also a higher frequency of lower lobe disease than is usual with chronic pulmonary aspergillosis and aspergillomas. In one case (patient 4) the disease was more nodular, another relatively underreported manifestation of CPA.[20]

Any bullous or cavitary pulmonary disease may be complicated by CPA.[11] Occasionally, patients with acute or subacute invasive aspergillosis (otherwise termed *chronic necrotising pulmonary aspergillosis*) develop into CPA. Aspergillomas are found in about 25% of patients with CPA,[21] in-

cluding occasional simple (single) aspergillomas in a pre-existing cavity. The frequency of CPA following tuberculosis is 15–22% in patients with persistent pulmonary cavities,[11,22] and an estimation of the global five-year period prevalence is 1,174,000 of such patients, or 18/100,000 in the population.[3] However, as an isolated radiological finding on chest radiographs, aspergillomas are rare with a reported prevalence of 0.13%.[23]

The finding of multiple aspergillomas is therefore rare, which is reflected in the lack of associated literature. There were two early reports of individual patients with multiple aspergilloma.[24,25] When CT scanning was introduced, workers at the Brompton Hospital in London noted that 4 (16%) of 26 patients with aspergillomas had bilateral lesions.[18] High-resolution CT of the thorax may facilitate

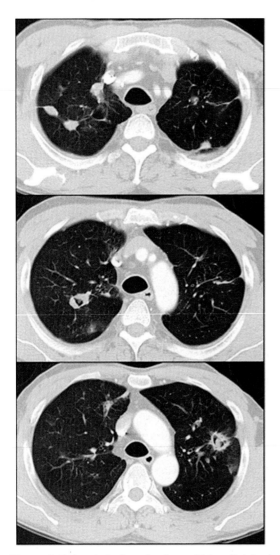

Figure 4. Patient 4. Sections showing multiple well-defined nodules in the upper lobes bilaterally. There were also two cavities inferiorly in the right upper and left upper lobes.

increased diagnosis. More recently four other reports[19,26–28] described bilateral aspergillomas. Shah *et al.* found six patients with multiple fungal balls occurring in the dilated and fusiform bronchial tree, among 41 patients undergoing autopsy or surgical resection.[29] However, little is written about the pathogenesis, diagnosis, and management of this complex presentation. The events that lead up to and result in multiple aspergilloma are yet to be described. This may be in part due to the often-late presentation of aspergilloma, with patients being largely asymptomatic in the earlier stages of the disease.

Aspergillomas have been classified as simple and complex. Simple aspergillomas have smooth edges and sit in a thin-walled cavity. Aspergillomas in a thick-walled cavity, often with surrounding inflammatory disease, have been termed *complex aspergilloma*,[30,31] although this term has also been broadened to include multicavity disease (often containing one or more aspergillomas). Patients 1, 2, and 5 above clearly have multiple, simple aspergillomas, in the sense that they are discrete and thin walled. Singh *et al.*[23] noted that only 10% of their aspergillomas were thin walled, a figure rarely quoted in series. Perhaps these thin-walled cavities have arisen within bronchi causing massive bronchial dilatation, with almost no parenchymal reaction? This is what appears to have occurred in case 3 of Lageze *et al.*,[32] as the presence of multiple bronchi entering and leaving the large cavity suggests this. Such appearances are in stark contrast to those described by the majority of other reports, although Shah's patients might have had a similar disease in some instances.

Patient 2 above may set the record for the largest number of aspergillomas ($n = 11$), with up to six having been seen previously.[33] The architecture of the lungs of patient 2 was one of gross multiple thin-walled cavitation, notably in the upper lobes. Aspergilloma occurrence in CF was reported in three patients from Dublin,[34] and it is likely that most of these fungal balls arose within bronchi. This would be the most likely explanation for the multiple lesions in patient 2, and possibly accounts for the thin walls of each aspergilloma and lack of pleural involvement. As there was so little parenchymal and pleural involvement, the transplant option was deemed best for her, given her dismal overall lung function. Rupture of a cavity with a pneumothorax (patient 3) is also a rare event,[35–37] although this could be the pathogenesis for *Aspergillus* empyema.

Multiple nodules that do not display a thin-walled cavity, or the characteristic crescent sign of an aspergilloma, were also seen in the right upper lobe of patient 4, which may represent another manifestation of chronic pulmonary aspergillosis.[14] Such nodular lesions vary in size and can be cavitating or not, but crucially they also display fungal hyphae on biopsy. While malignancy and other nodular radiological appearances should not be excluded, this presentation should be further examined in CPA patients. These lesions may evolve into

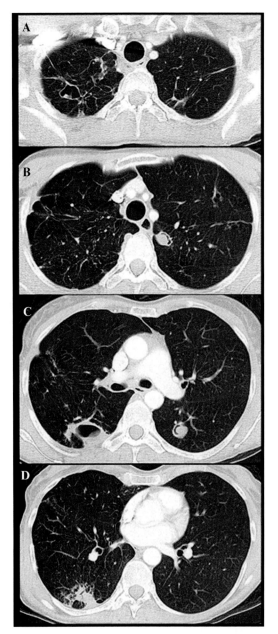

Figure 5. Patient 5. Four CT thorax sections in September 2010 showing four cavities, two containing aspergillomas. The highest section (A) shows a right sided thin-walled cavity with a small amount of material probably adherent posteriorly to the wall. (B) A second cavity on the left side medially almost completely filled with a fungal ball and an air crescent sign. (C) Two further cavities, one on each side, with that on the left with a slightly thicker wall and also almost completely filled with a fungal ball, with an air crescent sign. On the right, posteriorly abutting the pleura, is a thick-walled cavity with associated pleural thickening and two small areas of consolidation laterally. (D) The right-sided cavity is shown to extend several centimeters inferiorly, and change its shape slightly, but without intracavitary material.

cavities with or without fungal balls. Prior to this, lung cancer is the major diagnostic consideration, and surgical resection or biopsy is appropriate to make a histological diagnosis of aspergillosis.[38] Imaging techniques such as PET scanning have also demonstrated the ambiguous nature of an aspergilloma radiologically.[16] Further longitudinal study is required to understand the development of this form of aspergillosis. Other fungal differential diagnoses for pulmonary nodules in nonimmunocompromised patients include coccidioidomycosis[39] and cryptococcosis.[40]

Aspergillus precipitins (IgG antibody) are detectable in over 95% of patients with aspergilloma.[41] About 40% of patients are sensitized to *A. fumigatus* by specific IgE and skin prick testing.[42] Serology was used in all patients to aid diagnosis of aspergilloma or fungal infection. Patient 3 had only *Aspergillus* IgE antibodies detectable, an unusual serological pattern.

While approximately 10% of aspergillomas are reported to lyse spontaneously,[22,43] in the context of multiple lesions this is probably irrelevant to the progression of disease. Most patients with CPA have a progressive downward course, with general ill health including weight loss and fatigue, many pulmonary symptoms of breathlessness, productive cough, chest discomfort, and hemoptysis. In the first six months after diagnosis, 30% succumb to this and the closely related subacute invasive aspergillosis, often with multiple contributory factors.[44] Hemoptysis and pneumonia are the common causes of death. Corticosteroids may provide temporary relief and improvement in general symptoms but ultimately lead to progression of infection,[8,45] as demonstrated in several patients in this series. Long-term medical management is clearly required for these patients as surgery has a limited role with multiple lesions, though the emergence of antifungal resistance in three patients (demonstrated in patients 2 and 4 and inferred in patient 1) is a concern. Azole resistance is an emerging problem in the context of allergic and chronic aspergillosis.[3,16,21,46]

In summary, we have analyzed the characteristics of patients with multiple aspergillomas, which is a rare manifestation of chronic pulmonary aspergillosis. A single aspergilloma is typcially seen radiologically as a spherical mass in a thin-walled cavity displaying a crescent of air or cavitating nodules. Management is challenging and needs additional study.

Conflicts of interest

The authors declare no conflicts of interest.

References

1. Mullins, J. & A. Seaton. 1978. Fungal spores in lung and sputum. *Clin. Allergy* [Internet]. **8:** 525–533. Available from: http://www.ncbi.nlm.nih.gov/pubmed/361283. Accessed October 25, 2011.

2. Lass-Flörl, C., G. Kofler, G. Kropshofer, *et al.* 1998. In-vitro testing of susceptibility to amphotericin B is a reliable predictor of clinical outcome in invasive aspergillosis. *J. Antimicrob. Chemother.* **42:** 497–502. Available from: http://www.ncbi.nlm.nih.gov/pubmed/9818749. Accessed October 25, 2011.

3. Denning, D.W., A. Pleuvry & D.C. Cole. 2011. Global burden of chronic pulmonary aspergillosis as a sequel to pulmonary tuberculosis. *World Health* **89:** 1–21. Available from www.aspergillus.org.uk (library). Accessed October 13, 2012.

4. McCarthy, D.S. & J. Pepys. 1971. Allergic bronchopulmonary aspergillosis. Clinical immunology: 2. Skin, nasal and bronchial tests. *Clin. Allergy* **1:** 415–432. Available from: http://www.ncbi.nlm.nih.gov/pubmed/4950529. Accessed October 25, 2011.

5. Chakrabarti, A., S. Sethi, D.S.V Raman & D. Behera. 2002. Eight-year study of allergic bronchopulmonary aspergillosis in an Indian teaching hospital. *Mycoses* **45:** 295–299. Available from: http://www.ncbi.nlm.nih.gov/pubmed/12572718. Accessed October 25, 2011.

6. Fairs, A., J. Agbetile, B. Hargadon, *et al.* 2010. IgE sensitization to Aspergillus fumigatus is associated with reduced lung function in asthma. *Am. J. Respir. Crit. Care Med.* **182:** 1362–1368. Available from: http://www.pubmedcentral.nih.gov/articlerender.fcgi?artid=3029929 &tool=pmcentrez&rendertype=abstract. Accessed October 25, 2011.

7. Bennett, J. 1842. On the parasitic vegetable structures found growing in living animals. *Trans. Roy. Soc. Edinburgh* **15:** 167–174. Available from: http://www.aspergillus.org.uk/secure/treatment/aspllomref .php.

8. Denning, D.W., K. Riniotis, R. Dobrashian & H. Sambatakou. 2003. Chronic cavitary and fibrosing pulmonary and pleural aspergillosis: case series, proposed nomenclature change, and review. *Clin. Infect. Dis.* **37**(Suppl 3): S265–S280. Available from: http://www.ncbi.nlm.nih.gov/pubmed/12975754.

9. Camuset, J, A. Lavolé, M. Wislez, *et al.* 2007. Bronchopulmonary aspergillosis infections in the nonimmunocompromised patient. *Revue de Pneumologie Clinique* **63:** 155–166. Available from: http://www.ncbi.nlm.nih.gov/pubmed/17675939. Accessed October 25, 2011.

10. Kohno, S, K. Izumikawa, K. Ogawa, *et al.* 2010. Intravenous micafungin versus voriconazole for chronic pulmonary aspergillosis: a multicenter trial in Japan. *J. Infect.* **61:** 410–418. Available from: http://www.ncbi.nlm.nih.gov/pubmed/20797407. Accessed November 2, 2011.

11. Smith, N.L. & D.W. Denning. 2011. Underlying conditions in chronic pulmonary aspergillosis, including simple aspergilloma. *Eur. Respir. J.* **44:** 1–8. Available from: http://www.ncbi.nlm.nih.gov/pubmed/20595150. Accessed January 18, 2011.

12. Deve, F. 1938. Une nouvelle forme anatomo-radiologique de mycose pulmonaire primitive, le mega-mycetoma intra bronchiectasique. *Arch. Med. Chir de l'app Respir.* 13: 337–361. Available from www.aspergillus.org.uk (history). Accessed October 13, 2012.

13. Denning, D.W. 2001. Chronic forms of pulmonary aspergillosis. *Clin. Microbiol. Infect.* **7**(Suppl 2): 25–31. Available from: http://www.ncbi.nlm.nih.gov/pubmed/11525215.

14. Jain, L.R. & D.W. Denning. 2006. The efficacy and tolerability of voriconazole in the treatment of chronic cavitary pulmonary aspergillosis. *J. Infect.* **52:** e133–e137. Available from: http://www.ncbi.nlm.nih.gov/pubmed/16427702. Accessed March 11, 2012.

15. Denning, D.W., S. Park, C. Lass-Florl, *et al.* 2011. High-frequency triazole resistance found in nonculturable Aspergillus fumigatus from lungs of patients with chronic fungal disease. *Clin. Infect. Dis.* **52:** 1123–1129. Available from: http://www.pubmedcentral.nih.gov/articlerender.fcgi?artid=3106268&tool=pmcentrez&rendertype=abstract. Accessed July 19, 2011.

16. Howard, S.J., D. Cerar, M. Anderson, *et al.* 2009. Frequency and evolution of azole resistance in aspergillus fumigatus associated with treatment failure. *Emerging Infect. Dis.* **15:** 1068–1076. Available from: http://www.pubmedcentral.nih.gov/articlerender.fcgi?artid=2744247&tool=pmcentrez&rendertype=abstract. Accessed July 13, 2011.

17. Hope, W.W., E.M. Billaud, J. Lestner & D.W. Denning. 2008. Therapeutic drug monitoring for triazoles. *Curr. Opin. Infect. Dis.* **21:** 580–586. Available from: http://www.ncbi.nlm.nih.gov/pubmed/18978525. Accessed March 22, 2012.

18. Roberts, C.M., K.M. Citron & B. Strickland. 1987. Intrathoracic Aspergilloma: role of CT in diagnosis and treatment. *Radiology* **165:** 123. Available from: http://radiology.rsna.org/content/165/1/123.short. Accessed February 16, 2011.

19. Shah, A. 2008. Aspergillus-associated hypersensitivity respiratory disorders. *Indian J. Chest Dis. Allied Sci.* **50:** 117–128. Available from: http://www.ncbi.nlm.nih.gov/pubmed/18610696.

20. Baxter, C.G., P. Bishop, S.E. Low, *et al.* 2011. Pulmonary aspergillosis: an alternative diagnosis to lung cancer after positive [18F]FDG positron emission tomography. *Thorax* **66:** 638–640. Available from: http://www.ncbi.nlm.nih.gov/pubmed/21460371. Accessed Novembr 2, 2011.

21. Felton, T.W., C. Baxter, C.B. Moore, *et al.* 2010. Efficacy and safety of posaconazole for chronic pulmonary aspergillosis. *Clin. Infect. Dis.* **51:** 1383–1391. Available from: http://www.ncbi.nlm.nih.gov/pubmed/21054179. Accessed January 18 2011.

22. British Thoracic & Tuberculosis Association, T. 1970. Aspergilloma and residual tuberculous cavities - the results of a resurvey. *Tubercle.* **51:** 227–245. Available from www.aspergillus.org.uk (library). Accessed October 31, 2012.

23. Singh, P, P. Kumar, R. Bhagi & R. Singla. 1989. Pulmonary aspergilloma. *Indian J. Chest Dis. Allied Sci.* **31:** 177–185. Available from: http://thorax.bmj.com/content/ 38/8/572.abstract. Accessed October 23, 2011.

24. Le Nouene, Esquirol, Ardaillou. 1956. Aspergillome Bronchiectasiant Multiple (3 Localisations). *La Presse Medicale* **64:** 974–976. Available from www.aspergillus.org.uk (history). Accessed October 13, 2012.

25. Personne, C. 1961. [A henceforth common problem of pneumology: pulmonary aspergilloma]. *Rev. Tuberc Pneumol–(Paris)* **25:** 1434–1448. Available from www.aspergillus.org.uk (history). Accessed October 13, 2012.

26. Aderaye, G & A. Jajaw. 1996. Bilateral pulmonary aspergilloma: case report. *East African Med. J.* **73:** 487–488. Available from: http://www.ncbi.nlm.nih.gov/pubmed/ 8918017. Accessed October 25, 2011.

27. Howard, S.J., I. Webster, C.B. Moore, *et al.* 2006. Multi-azole resistance in Aspergillus fumigatus. *Int. J. Antimicrob. Agents* **28:** 450–453.

28. Campbell, N, H. Stunell, S. Barrett & W.C. Torreggiani. 2009. Multiple pulmonary aspergillomas. *JBR-BTR: organe de la Société royale belge de radiologie (SRBR)=orgaan van de Koninklijke Belgische Vereniging voor Radiologie (KBVR)* **92:** 118. Available from: http://www.ncbi. nlm.nih.gov/pubmed/19534248. Accessed Oct 25, 2011.

29. Shah, R, P. Vaideeswar & S.P. Pandit. 2008. Pathology of pulmonary aspergillomas. *Indian J. Pathol. Microbiol.* **51:** 342–345. Available from: http://www.ncbi.nlm.nih. gov/pubmed/18723954. Accessed October 25, 2011.

30. Ahmad, T., S.W. Ahmed, N. Hussain & K. Rais. 2010. Clinical profile and postoperative outcome in patients with simple and complex aspergilloma of lung. *J. Coll. Phys. Surg.–Pakistan* **20:** 190–193. Available from: http://www.ncbi. nlm.nih.gov/pubmed/20392383.

31. Sales, M. & P.U. da. 2009. Continuing education course—Mycoses Chapter 5—Aspergillosis: from diagnosis to treatment. *J. Bras. Pneumol.* **35:** 1238–1244.

32. Lageze, P, R. Touraine & R. Patin. 1953. [Development of pulmonary aspergilloma in a fibrous tuberculous cavern]. *Lyon Méd.* **189:** 132–138. Available from: http://www.ncbi.nlm.nih.gov/pubmed/13098319. Accessed October 27, 2011.

33. Plihal, V, Z. Jedlickova, J. Viklicky & A. Tomanek. 1964. Multiple Bilateral Pulmonary Aspergillomata. *Thorax* 19: 104–111. Available from: http://www.pubmedcentral.nih. gov/articlerender.fcgi?artid=1018807&tool=pmcentrez& rendertype=abstract.

34. Maguire, C.P., J.P. Hayes, M. Hayes, *et al.* 1995. Three cases of pulmonary aspergilloma in adult patients with cystic fibrosis. *Thorax* **50:** 805–806. Available from: http://www.pubmedcentral.nih.gov/articlerender.fcgi?artid =474659&tool=pmcentrez&rendertype=abstract. Accessed October 25, 2011.

35. Baradkar, V.P., M. Mathur & S. Kumar. 2009. Uncommon presentation of pulmonary aspergilloma. *Indian J. Med. Microbiol.* **27:** 270–272. Available from: http://www.ncbi. nlm.nih.gov/pubmed/19584515. Accessed January 27, 2011.

36. Sakuraba, M, Y. Sakao, A. Yamazaki, *et al.* 2006. A case of aspergilloma detected after surgery for pneumothorax. *Ann. Thor. Cardiovasc. Surg.* **12:** 267–269. Available from: http://www.ncbi.nlm.nih.gov/pubmed/16977297. Accessed October 25, 2011.

37. Wood, J.R., D. Bellamy, A.H. Child & K.M. Citron. 1984. Pulmonary disease in patients with Marfan syndrome. *Thorax* **39:** 780–784. Available from: http://www.pubmedcentral.nih.gov/articlerender.fcgi?artid =459918&tool=pmcentrez&rendertype=abstract. Accessed October 25, 2011.

38. Park, Y., T.S. Kim, C.A. Yi, *et al.* 2007. Pulmonary cavitary mass containing a mural nodule: differential diagnosis between intracavitary aspergilloma and cavitating lung cancer on contrast-enhanced computed tomography. *Clin. Radiol.* **62:** 227–232. Available from: http://www.ncbi. nlm.nih.gov/pubmed/17293215. Accessed January 28, 2011.

39. Parish, J.M. & J.E. Blair. 2008. Coccidioidomycosis. Mayo Clinic proceedings. *Mayo Clin.* **83:** 343–348; quiz 348–349. Available from: http://www.ncbi.nlm.nih.gov/ pubmed/18316002. Accessed July 24, 2010.

40. Ou, K.-W., K.-F. Hsu, Y.-L. Cheng, *et al.* 2010. Asymptomatic pulmonary nodules in a patient with early-stage breast cancer: cryptococcus infection. *Int. J. Infect. Dis.* **14:** e77–e80. Available from: http://www.ncbi. nlm.nih.gov/pubmed/19477671. Accessed February 2, 2011.

41. Kurup, V.P. & A. Kumar. 1991. Immunodiagnosis of aspergillosis. *Clin. Microbiol. Rev.* **4:** 439–456. Available from: http://www.pubmedcentral.nih.gov/articlerender.fcgi?artid =358211&tool=pmcentrez&rendertype=abstract. Accessed February 8, 2011.

42. Denning, D.W., B.R. O'Driscoll, G. Powell, *et al.* 2009. Randomized controlled trial of oral antifungal treatment for severe asthma with fungal sensitization: the Fungal Asthma Sensitization Trial (FAST) study. *Am. J. Respir. Crit. Care Med.* **179:** 11–18. Available from: http://www.ncbi.nlm.nih.gov/pubmed/18948425. Accessed June 30, 2010.

43. Hammerman, K.J., C.S. Christianson, I. Huntington, *et al.* 1973. Spontaneous lysis of aspergillomata. *Chest* **64:** 697–699. Available from: http://www.chestjournal. org/cgi/doi/10.1378/chest.64.6.697. Accessed January 27, 2011.

44. Nam, H.-S., K. Jeon, S.-W. Um, *et al.* 2010. Clinical characteristics and treatment outcomes of chronic necrotizing pulmonary aspergillosis: a review of 43 cases. *Int. J. Infect. Dis.* **14:** e479–e482. Available from: http://www.ncbi.nlm.nih.gov/pubmed/19910234. Accessed July 13, 2011.

45. Davies, D. & R.A. Somner. 1972. Pulmonary aspergillomas treated with corticosteroids. *Thorax* **27:** 156–162. Available from: http://www.pubmedcentral.nih. gov/articlerender.fcgi?artid=472514&tool=pmcentrez& rendertype=abstract.

46. Mortensen, K. & R. Jensen. 2011. Aspergillus species and other molds in respiratory samples from patients with cystic fibrosis: a laboratory-based study with focus on aspergillus fumigatus azole resistance. *J. Clin. Microbiol.* **49:** 2243–2251. Available from: http://jcm.asm.org/content/ 49/6/2243.short. Accessed March 23, 2012.

Ann. N.Y. Acad. Sci. ISSN 0077-8923

ANNALS OF THE NEW YORK ACADEMY OF SCIENCES
Issue: *Advances Against Aspergillosis*

Immune regulation in idiopathic bronchiectasis

Rosemary J. Boyton and Daniel M. Altmann

Department of Medicine, Section of Infectious Diseases and Immunity, Lung Immunology Group and Human Disease Immunogenetics Group, Imperial College, London, United Kingdom; and Department of Respiratory Medicine, Royal Brompton & Harefield NHS Foundation Trust, London, United Kingdom

Address for correspondence: Dr. Rosemary J. Boyton, Department of Medicine, Section of Infectious Diseases and Immunity, Lung Immunology Group and Human Disease Immunogenetics Group, Commonwealth Building 8N22, Hammersmith Campus, Du Cane Road, Imperial College, London W12 ONN, UK. r.boyton@imperial.ac.uk

Bronchiectasis is a complex pathological endpoint arrived at through a diverse interplay between lung infection and altered immune function. It comprises irreversible, abnormal dilatation of one or more bronchi, with chronic airway inflammation and is associated with recurrent chest infections, airflow obstruction, chronic cough, excessive sputum production, and malaise. Many pathogens are associated with this disease, including chronic bacterial infections, nontuberculous mycobacteria, and aspergillis. However, the etiology is poorly defined. Disease-associated genes indicate a likely contribution to disease mechanism both from innate and adaptive immunity. The role of immune mechanisms is highlighted by the occurrence of bronchiectasis in a subset of patients with rheumatoid arthritis or inflammatory bowel disease as well as diseases of immune dysregulation such as combined variable immune deficiency, transporter associated with antigen processing (TAP) deficiency syndrome, and hyper-IgE syndrome. Recent evidence indicates a possible role of excessive natural killer cell activation in pathogenesis.

Keywords: bronchiectasis; chronic lung infection; allergic bronchopulmonary aspergillosis; NK cells; aspergillus; immunity

Bronchiectasis is commonly defined as an irreversible, abnormal dilatation of one or more bronchi, with chronic airway inflammation.[1] This is associated with recurrent chest infections, airflow obstruction, chronic cough, excessive sputum production, and malaise. However, the underlying etiology is poorly defined, and these may best be regarded as common clinical and pathological endpoints, potentially resulting from diverse underlying causes. Evidence suggests that during the latter part of the 20th century, when the incidence of bronchiectasis might have been expected to fall due to changes in immunization and antibiotic prescribing, the disease became more common—an increase probably not attributable simply to improved detection.[2] It had been thought that the prevalence of bronchiectasis was decreasing in the developed world as a result of childhood immunizations, increased use of antibiotics, and decreased tuberculosis. A U.S. retrospective study for the period 1993–2006 showed an average age-adjusted annual increase in hospitalizations of 2.4% in men and 3.0% in women.[2] The explanation for this apparent paradox may lie in the assumption that bronchiectasis should be regarded as neither a simple susceptibility to chronic infection nor as a physical defect of the bronchi, but as a complex endpoint of the interplay between microbial pathogens, host microbiota, and the influence of these on host regulation of inflammation and adaptive immunity.

Different populations show considerable differences in prevalence of the disease, an observation most likely related to differences both in genetics and environment, as discussed later. In Australian Aborigines[3] and Alaskan native children, the prevalence is 14 per 1,000, compared with around 1 per 1,000,000 in Finnish children.[4]

The majority of bronchiectasis cases are described as idiopathic insofar as no causal, initiating factor is identifiable. Other causes include postinfectious, immune deficiency–associated, and allergic bronchopulmonary aspergillosis (ABPA).[1] There is also

doi: 10.1111/j.1749-6632.2012.06756.x

an interesting subset of patients in which bronchiectasis is associated with autoimmune disease, most notably, rheumatoid arthritis and ulcerative colitis. This list of associated, predisposing factors exemplifies the mechanistic conundrum of this disease process: since the work of Peter Cole in the 1980s, it has been argued that this should be considered the result of a vicious cycle of chronic infection and dysregulated, excessive inflammation.[5] Certainly, a key feature, both of bronchiectasis and ABPA, is the chronic interleukin (IL)-8–mediated homing of neutrophils to the lung.[6,7] It is also possible that neutrophil migration is promoted by IL-17 release from lung T cells, gamma delta ($\gamma\delta$)) cells, and natural killer (NK) cells responding to local bacterial infection. However, this is a disease endpoint that may be arrived at either by deficiencies in immunity (notably immunoglobulin deficiencies and transporter associated with antigen processing (TAP) deficiency syndrome, in which there is abrogated TAP function necessary for the proper expression of HLA class I molecules loaded with antigenic peptide)[8] or by conditions that might be regarded as comprising excessive, dysregulated immune effector mechanisms, such as rheumatoid arthritis and ulcerative colitis. Although bronchiectasis is evidently not primarily autoimmune in etiology, there are important clues to an underlying mechanism offered by the appearance of these symptoms in a proportion of patients with these autoimmune conditions.[9,10] Furthermore, combined variable immune deficiency (CVID) patients are prone to both bronchiectasis and to a wide range of autoimmune pathologies.[11]

It is interesting to consider the role of *Aspergillus fumigatus* as a trigger for bronchiectasis in the context of this spectrum of immune competence. Infection with aspergillus may elicit diverse host Th1/Th2/Th17 subset polarization depending on several factors, including immunogenetic background and metabolic status of the fungal spores themselves.[12,13] In the context of immune hyperreactivity, lung exposure to fungal spores may lead to immunopathologically mediated ABPA. However, impaired immunity will predispose to invasive aspergillosis with pathology dependent on fungal enzymes. In general, patients with chronic lung disease show enhanced susceptibility to *A. fumigatus* colonization and infection.[14]

Hyper-IgE syndrome, or Job's syndrome, an immune deficiency resulting from mutations in the STAT3 cytokine signaling transducer, is associated with raised IgE, highly impaired Th17 immunity, and an elevated proinflammatory profile with respect to tumor necrosis factor-alpha (TNF-α) and interferon-gamma (IFN-γ).[16,17] Furthermore, it has been proposed that defective generation of IL-10–induced tolerogenic dendritic cells and inducible regulatory T cells may contribute to proinflammatory changes.[18] This immune defect underlines the fact that bronchiectasis can be regarded not simply as an outcome from insufficient immunity but from dysregulated control of the innate and adaptive response to infection. Around two-thirds of these patients manifest bronchiectasis among a range of clinical symptoms that also includes dermatitis, mucocutaneous infections due to *Staphylococcus aureus* and *Candida,* as well as pneumonia, mostly caused by *S. aureus* or *Streptococcus pneumoniae.*[15] Around one half of hyper-IgE patients with structural lung disease have invasive fungal disease.[19]

Bronchiectasis is generally associated with increased susceptibility to a number of specific pathogens including *Haemophilus influenza, H. parainfluenzae, Pseudomonas aeruginosa, S. pneumoniae, Moraxella catarrhalis, S. aureus, Stenotrophomonas maltophilia,* Gram-negative enterobacter, nontuberculosis mycobacteria (NTM), and aspergillus. The prevalence of NTM in patients with bronchiectasis is 2%, with mycobacterium avium complex (MAC) the most frequent NTM isolated. *P. aeruginosa* and *S. aureus* are frequently cocultured. Bronchiectasis patients with NTM show a higher prevalence of coexisting aspergillus-related lung disease.[20]

Genome-wide association studies have not been undertaken for idiopathic bronchiectasis, but a number of candidate gene polymorphisms associated with innate and adaptive immunity have been investigated. Although no association has been found with functional polymorphisms for TLR4 or TLR2,[21] an IFN-γ polymorphism and a CXCR-1 polymorphism, which influences IL-8 binding and neutrophil trafficking, were strongly associated with bronchiectasis risk following colectomy in ulcerative colitis, though not with enhanced bronchiectasis risk per se.[22] HLA class II analysis shows an association with HLA-DRB1*0101 and HLA-DQB1*0501, both commonly present in linkage disequilibrium in the same DR1 haplotype.[23] This might be interpreted to indicate a classical immune response gene effect with respect to adaptive immunity to a key microbial pathogen involved in triggering disease. In line with this notion, CD4[+] T cells are indeed found in lung biopsy tissue from patients, although CD8[+] T cells are if anything more abundant.[24] This presumably indicates an additional, common component to microbial pathogenesis that has not yet been substantially investigated, whether viral in nature or dependent on intracellular bacteria.

Potential roles for other cell types in immunopathogenesis are indicated by more indirect evidence. The rare individuals with TAP deficiency syndrome suffer from a much more specific spectrum of infections and pathologies than might perhaps be predicted for a defect as fundamental to adaptive immunity as the inability to load antigenic peptides into the peptide-binding groove of HLA class I molecules for presentation to T cells. Within this spectrum of pathologies is bronchiectasis.[8] It has been suggested that this and other defects in mucosal immunity in these families may be a consequence of abnormal activation of NK cells or γδ T cells, downstream of the HLA class I loading defect.

Alterations in NK cell activation are currently a source of intense analysis in a very wide range of conditions, spanning tumors, infections, autoimmunity, and inflammation. NK cells were once considered simply "null" cells—lymphocytes lacking the specific antigen receptors of either B cells or T cells and yet capable of spontaneously lysing tumors or herpesvirus-infected targets.[25,26] This view was followed by a period in which a model was developed to account for experiments showing that NK cell activation ensued from recognition of cells through downregulated expression of HLA class I molecules.[27] In the past 20 years, it has become clear that NK cells follow a very complex program of counterpoised receptor families that impart an integrated array of activating and inhibitory signals to the nucleus.[28]

Furthermore, NK cells are likely to be as diverse as CD4[+] T cells in terms of cellular subsets, expressing different receptors and differentiation markers and follow different transcriptional programs, from proinflammatory to anti-inflammatory, in addition to effects that feed into subsequent polarization of adaptive T cells immunity. The NK cell paradigm has also shifted to encompass the fact that, far from being limited to recognition of tumors and herpesviruses, they haves roles in immunity to autoantigens as well as to bacterial, viral, fungal, and parasite antigens.

NK cells are considered to be extremely important in the normal functioning of lung host defense.[29,30] Current models of NK cell activation have focused in part on the component of activation dependent on the interaction between HLA class I molecules and a large family of cognate receptors— the killer immunoglobulin-like receptors (KIR). The KIR genes are encoded on human chromosome 19q13.4 as a string of genes of variable number and identity between individuals. Depending on the signaling domains attached to their cytoplasmic tails, an immunoreceptor tyrosine-based inhibition motif (ITIM) or an immunoreceptor tyrosine-based activation motif (ITAM), KIRs may impart an inhibitory or activating signal, respectively. This is an extreme form of immune diversity within and between populations: there is considerable variation of geography and ethnicity in the composition of KIR haplotypes.[23] The significance of the interesting observation that some populations, such as aboriginal Australians who have the highest prevalence of bronchiectasis, also tend to carry some of the most activating KIR haplotypes has yet to be elucidated. There is further diversity and complexity due to the fact that different individuals within a given population will carry different numbers of activating and inhibitory KIR genes and different polymorphic variants of these genes.[31] Furthermore, the impact of the KIR repertoires will vary depending on whether a given individual also carries the cognate HLA class I polymorphic variant with which the KIRs interact. Finally, there is variegated expression of KIR gene products, such that they are differentially expressed on the NK cells (and T cells) within an individual. In analysis of our UK idiopathic bronchiectasis cohort, HLA-C group 1 homozygosity, specifically, the HLA-Cw∗03 allele, was associated with disease.[23,32] Around 50% of this cohort was HLA-C group 1 homozygous, compared with 25% of controls. It would be predicted that fewer NK cells would fall under inhibitory control if the ligands for inhibitory receptors are absent. In bronchiectasis, there is a preponderance of only HLA-C group 1 with the activating KIRs 2DS1 and/or 2DS2. Such combinations might be predicted to contribute to a highly activated (and potentially proinflammatory) NK cell program.[32]

Concluding remarks

Bronchiectasis is a pathological endpoint resulting from a complex interplay between lung infection and immunity. The interaction both with diverse forms of immune deficiency of distinct immune pathways and with a subset of autoimmune patients offers particular conceptual challenges. Recent immunogenetic analysis, predicting a pathogenic role for excessive NK cell activation, may offer new therapeutic options.

Conflicts of interest

The authors declare no conflicts of interest.

References

1. Boyton, R.J. 2012. Bronchiectasis. *Medicine* **40**: 5, 267–272.
2. Seitz, A.E., K.N. Olivier, C.A. Steiner, *et al.* 2010. Trends and burden of bronchiectasis-associated hospitalizations in the United States, 1993-2006. *Chest* **138**: 944–949.
3. Chang, A.B., K. Grimwood, E.K. Mulholland & P.J. Torzillo. 2002. Working group on indigenous paediatric respiratory health. Bronchiectasis in indigenous children in remote Australian communities. *Med. J. Aust.* **177**: 200–204.
4. Säynäjäkangas, O., T. Keistinen, T. Tuuponen & S.L. Kivelä. 1998. Evaluation of the incidence and age distribution of bronchiectasis from the Finnish hospital discharge register. *Cent. Eur. J. Public Health* **6**: 235–237.
5. Cole, P. 1984. A new look at the pathogenesis and management of persis- tent bronchial sepsis: a "vicious circle" hypothesis and its logic.al therapeutical connotations. In *Strategies for the Management of Chronic Bronchial Sepsis.* R.J. Davies, Ed. 1–20. The Medicine Publishing Foundation. Oxford.
6. Gibson, P.G., P.A. Wark, J.L. Simpson, *et al.* 2003. Induced sputum IL-8 gene expression, neutrophil influx and MMP-9 in allergic bronchopulmonary aspergillosis. *Eur. Respir. J.* **21**: 582–588.
7. Gaga, M., A.M. Bentley, M. Humbert, *et al.* 1998. Increases in CD4+ T lymphocytes, macrophages, neutrophils and interleukin 8 positive cells in the airways of patients with bronchiectasis. *Thorax* **53**: 685–691.
8. Gadola, S.D., H.T. Moins-Teisserenc, J. Trowsdale, *et al.* 2000. TAP deficiency syndrome. *Clin. Exp. Immunol.* **121**: 173–178.
9. Tsuchiya, Y., N. Takayanagi, H. Sugiura, *et al.* 2011. Lung diseases directly associated with rheumatoid arthritis and their relationship to outcome. *Eur. Respir. J.* **37**: 1411–1417.
10. Wilczynska, M.M., A.M. Condliffe & D.J. McKeon. 2012. Coexistence of bronchiectasis and rheumatoid arthritis: revisited. *Respir. Care.* doi: 10.4187/respcare.01857. July 10 [Epub ahead of print].
11. Maarschalk-Ellerbroek, L.J., A.I. Hoepelman, J.M. van Montfrans & P.M. Ellerbroek. 2012. The spectrum of disease manifestations in patients with common variable immunodeficiency disorders and partial antibody deficiency in a university hospital. *J. Clin. Immunol.* **32**: 907–921.
12. Rivera, A., H.L. Van Epps, T.M. Hohl, *et al.* 2005. Distinct CD4+-T-cell responses to live and heat-inactivated Aspergillus fumigatus conidia. *Infect. Immun.* **73**: 7170–7179.
13. Rivera, A., T.M. Hohl, N. Collins, *et al.* 2011. Dectin-1 diversifies aspergillus fumigatus-specific T cell responses by inhibiting T helper type 1 CD4 T cell differentiation. *J. Exp. Med.* **208**: 369–381.
14. Böllert, F.G., P.J. Sime, W. MacNee & G.K. Crompton. 1994. Pulmonary mycobacterium malmoense and aspergillus infection: a fatal combination? *Thorax* **49**: 521–522.
15. Chandesris, M.O., I. Melki, A. Natividad, *et al.* 2012. Autosomal dominant STAT3 deficiency and Hyper-IgE syndrome: molecular, cellular, and clinical features from a French National Survey. *Medicine (Baltimore)* **91**: e1–e19.
16. Paulson, M.L., A.F. Freeman & S.M. Holland. 2008. Hyper IgE syndrome: an update on clinical aspects and the role of signal transducer and activator of transcription 3. *Curr. Opin. Allergy Clin. Immunol.* **8**: 527–533.
17. Milner, J.D., J.M. Brenchley, A. Laurence, *et al.* 2008. Impaired T(H)17 cell differentiation in subjects with autosomal dominant hyper-IgE syndrome. *Nature* **452**: 773–776.
18. Saito, M., M. Nagasawa, H. Takada, *et al.* 2011. Defective IL-10 signaling in hyper-IgE syndrome results in impaired generation of tolerogenic dendritic cells and induced regulatory T cells. *J. Exp. Med.* **208**: 235–249.
19. Vinh, D.C., J.A. Sugui, A.P. Hsu, *et al.* 2010. Invasive fungal disease in autosomal-dominanthyper-IgE syndrome. *J. Allergy Clin. Immunol.* **125**: 1389–1390.
20. Kunst, H., M. Wickremasinghe, A. Wells & R. Wilson. 2006. Nontuberculous mycobacterial disease and Aspergillus-related lung disease in bronchiectasis. *Eur. Respir. J.* **28**: 352–357.
21. Reynolds, C., L. Ozerovitch, R. Wilson, *et al.* 2007. Toll-like receptors 2 and 4 and innate immunity in neutrophilic asthma and idiopathic bronchiectasis. *Thorax* **62**: 279.
22. Boyton, R.J., C. Reynolds, F.N. Wahid, *et al.* 2006. IFN gamma and CXCR-1 gene polymorphisms in idiopathic bronchiectasis. *Tissue Antigens* **68**: 325–330.
23. Boyton, R.J. & D.M. Altmann. 2007. Natural killer cells, killer immunoglobulin-like receptors and human leucocyte antigen class I in disease. *Clin. Exp. Immunol.* **149**: 1–8.
24. Silva, J.R., J.A. Jones, P.J. Cole & L.W. Poulter. 1989. The immunological component of the cellular inflammatory infiltrate in bronchiectasis. *Thorax* **44**: 668–673.
25. Kiessling, R., G. Petranyi, K. Kärre, *et al.* 1976. Killer.cells: a functional comparison between natural, immune T-cell and antibody-dependent in vitro systems. *J. Exp. Med.* **143**: 772–780.
26. Glimcher, L., F.W. Shen & H. Cantor. 1977. Identification of a cell-surface antigen selectively expressed on the natural killer cell. *J. Exp. Med.* **145**: 1–9.
27. Ljunggren, H.G. & K. Kärre. 1990. In search of the 'missing-self': MHC molecules and NK cell recognition. *Immunol. Today* **11**: 237–244.
28. Cheent, K. & S.I. Khakoo. 2009. Natural killer cells: integrating diversity with function. *Immunology* **126**: 449–457.

29. Culley, F.J. 2009. Natural killer cells in infection and inflammation of the lung. *Immunology* **128:** 151–163.

30. Wang, J., F. Li, M. Zheng, *et al.* 2012. Lung natural killer cells in mice: phenotype and response to respiratory infection. *Immunology* **137:** 37–47.

31. Middleton, D. & F. Gonzelez. 2010. The extensive polymorphism of KIR genes. *Immunology* **129:** 8–19.

32. Boyton, R.J., J. Smith, R. Ward, *et al.* 2006. HLA-C and killer cell immunoglobulin-like receptor genes in idiopathic bronchiectasis. *Am. J. Respir. Crit. Care Med.* **173:** 327–333.

Ann. N.Y. Acad. Sci. ISSN 0077-8923

ANNALS OF THE NEW YORK ACADEMY OF SCIENCES
Issue: *Advances Against Aspergillosis*

Aspergillus bronchitis without significant immunocompromise

Ales Chrdle,[1]* Sahlawati Mustakim,[2] Rowland J. Bright-Thomas,[3] Caroline G. Baxter,[1,4] Timothy Felton,[1,4] and David W. Denning[1,4]

[1]The National Aspergillosis Center, University Hospital of South Manchester, Manchester, UK. [2]Pathology Department, Hospital Tengku Ampuan Rahimah, Klang, Selangor, Malaysia. [3]Cystic Fibrosis Unit, The University Hospital of South Manchester, Manchester, United Kingdom. [4]The University of Manchester, Manchester Academic Health Science Center, Manchester, United Kingdom

Address for correspondence: David W. Denning, Education and Research Center, University Hospital of South Manchester, Southmoor Road, Manchester, M23 9LT, United Kingdom. ddenning@manchester.ac.uk

Aspergillus bronchitis is poorly understood and described. We extracted clinical data from more than 400 referred patients with persistent chest symptoms who did not fulfill criteria for allergic, chronic, or invasive aspergillosis. Symptomatic patients with a positive culture or real-time PCR for *Aspergillus* spp. were reviewed. Seventeen patients fulfilled the selected criteria. Fourteen were women, with a mean age of 57 years (range 39–76). Sixteen of the patients had productive cough, eight had voluminous tenacious sputum, and seven had recurrent chest infections. Eight patients had Medical Research Council dyspnea scores of 4–5; 12 had bronchiectasis; and 13 patients grew *A. fumigatus*, 3 *A. niger*, and 1 *A. terreus*. Twelve of the 17 patients (71%) had elevated *Aspergillus* IgG (47–137 mg/L, mean 89.2) and 5 (29%) had elevated *Aspergillus* precipitins. Six of 12 (50%) had a major response to antifungal therapy and five of 12 (42%) patients relapsed, requiring long-term therapy. *Aspergillus* bronchitis is a discrete clinical entity in patients with structural lung disease but who are not significantly immunocompromised. It is distinct from asymptomatic fungal colonization and other forms of aspergillosis, and may respond to antifungal therapy.

Keywords: mannose binding lectin; fumigatus; niger; precipitins; bronchiectasis; aspergillary

Introduction

Aspergillus spp. exhibit a range of interactions with the human airway, including colonization, mucosal invasion, and provoking an allergic response.[1-3] *Aspergillus* infections that are limited entirely or predominantly confined to the tracheobronchial tree are currently termed *Aspergillus* trachcobronchitis[4] and usually occur in immunocompromised patients. Manifestations of *Aspergillus* tracheobronchitis include pseudomembranous tracheobronchitis, ulcerative tracheobronchitis (in lung transplant recipients), mucoid impaction of the bronchi (often in allergic bronchopulmonary aspergillosis [ABPA]), and obstructing bronchial aspergillosis.[5-7] Superficial tracheobronchitis is also recognized in lung transplant recipients.[6] Persistent airway colonization is characteristic of ABPA[8] and is sometimes seen in patients with chronic obstructive pulmonary disease (COPD)[9] and asthma.[10] Nearly forgotten is the entity *Aspergillus* bronchitis in non- or mildly immunocompromised patients.

The first description of *Aspergillus* tracheobronchitis was in 1890, in a nonimmunocompromised three-year-old girl who died[11] (Table 1) (Fig. 1). The first report, akin to the reports we offer here, was from Hoxie and Lamar in Kansas City writing in 1912 and described extreme coughing and husky voice affecting two patients with airways containing material consistent with *Aspergillus* spp.[12] Other notable descriptions include those of Wahl, Schneider, Von Ordstrand, and Riddell and Clayton (Table 1).[14,16,18,20] In 1962, Symmers described *Aspergillus* bronchitis from a pathological

*Current address: Tropical and Infectious Diseases Unit, Royal Liverpool University Hospital, Liverpool, UK

doi: 10.1111/j.1749-6632.2012.06816.x
Ann. N.Y. Acad. Sci. 1272 (2012) 73–85 © 2012 New York Academy of Sciences.

Table 1. Landmark papers with regard to *Aspergillus* tracheobronchitis and bronchitis

Date	No of cases	Description	Current terminology	Reference
1890	1	Autopsy description of a 3-year-old who died after a 10-day pneumonic illness of white adherent patches on the trachea and bronchi that showed microscopic features most consistent with *Aspergillus* spp.	Invasive *Aspergillus* tracheobronchitis	Wheaton[11]
1912	2	Two adults with cough and either breathlessness or husky voice had multiple sputa showing hyphae, one culture negative the other non-sporulating mold. Fungous tracheobronchitis.	*Aspergillus* bronchitis	Hoxie & Lamar[12]
1926	1	Presence of mucous membranes of the bronchi with ulceration; bronchitic aspergillosis	Pseudomembranous *Aspergillus* tracheobronchitis	Lapham[13]
1928	1	Severe symptoms of cough and wheeze following environmental exposure, with *A. flavus* cultured from sputum, which remitted after some months and NaI treatment	*Aspergillus* bronchitis	Wahl[14]
1928	0	Complete classification of all bronchopulmonary fungal infections.	Many	Castellani[15]
1930	1	Single case without underlying disease of cough and mild haemoptysis with low grade fever over 11 years with 5-year remission	*Aspergillus* bronchitis	Schneider[16]
1936	0	Review of the earlier work		Fawcitt[17]
1940	1	Three week illness with cough, fever, night sweats, weight loss in a nurseryman, who had no prior history. Sporulating *A. fumigatus* in sputum and positive skin prick test. Parenchymal infiltrates from both hilum.	Acute *Aspergillus* bronchitis	Van Ordstrand[18]
1952	0	Description of aspergilloma (including one bronchial aspergilloma) and probably ABPA. Detailed review of prior literature.	Bronchial aspergilloma and ABPA	Hinson[19]
1958	12	Patients with bronchitis who grew *Aspergillus* (usually *fumigatus*) in their sputum	*Aspergillus* bronchitis	Riddell & Clayton[20]
1962	>5	The first real description of the anatopathological features of *Aspergillus* bronchitis. Virtually saprophytic growth in the mixture of bronchial secretion, low grade inflammatory exudate and desquamated epithelial cells. Often conidiophores are found growing in the airways. Specially stained preparations "show very clearly that the fungus sometimes has a distinct attachment to the basement membrane (BM) of the chronically inflamed mucosa." The BM may become greatly thickened. Some of the hyphae end in single, vesicle-like expansions situated in contact with the thickened BM.	*Aspergillus* bronchitis	Symmers[21]
1964	35	All with *A. fumigatus* in their sputum, often heavy growths on many occasions. Long history of "bronchitis" with a productive cough. 25/35 (71%) had asthma. 0/28 were skin prick test positive. 11/35 (31%) had detectable *Aspergillus* precipitins. Aspergillary bronchitis.	*Aspergillus* bronchitis	Campbell & Clayton[22]

Continued

Table 1. *Continued*

Date	No of cases	Description	Current terminology	Reference
1970	8	Autopsy series showing *Aspergillus* bronchitis in 8 mildly immunocompromised patients described as a "localized form of aspergillosis characterized by bronchial casts containing mucus and mycelia." Invasion of mucosa is uncommon but there may be superficial erosion. Aspergillary bronchitis.	*Aspergillus* bronchitis and invasive *Aspergillus* tracheobronchitis	Young[23]
1989	2	Neutropenic patients with bronchial mucosa showing superficial erosions and ulcerations	Invasive *Aspergillus* tracheobronchitis	Berlinger & Freeman[24]
1991	6	Ulceration and invasion of cartilage in lung transplant recipients, especially around the anastomosis	Ulcerative *Aspergillus* tracheobronchitis	Kramer[5]
1991	3	HIV infected patients with obstruction of their airway with mucous containing abundant mucous, without ulceration or invasion	Obstructing (obstructive) *Aspergillus* tracheobronchitis	Denning[25]
1993	4	Severely immunocompromised AIDS, 3 with diffuse tracheobronchitis, multiple ulcerative or "plaque-like" inflammatory lesions, and occasionally nodules involving the mainstem and segmental bronchi. One patient had a single deep ulceration of the proximal trachea. There was variable invasion of the mucosa, submucosa, and cartilage and one had evidence of disseminated aspergillosis.	Invasive *Aspergillus* tracheobronchitis	Kemper[26]
1995	0	Review and classification with ulcerative tracheobronchitis, pseudomembranous tracheobronchitis, invasive tracheobronchitis, obstructive tracheobronchitis, and *Aspergillus* tracheobronchitis	All	Denning[4]
2005	1	Invasive pulmonary aspergillosis in AIDS, transformed by antifungal therapy and immune reconstitution into fatal obstructing bronchial aspergillosis	Obstructing *Aspergillus* tracheobrochitis	Sambatakou[27]
2006	6	*Aspergillus* bronchitis in cystic fibrosis patients growing *A. fumigatus* in sputum, who did not fulfill the criteria for ABPA, or respond to antibiotics, who made a good response to antifungal therapy	*Aspergillus* bronchitis	Shoseyov[28]

perspective.[21] He described *Aspergillus* growing as sporing mycelium with mild inflammation of the mucosa and/or excess production of mucus without invasion of the mucosa. Campbell and Clayton described the clinical features of "aspergillary" bronchitis as repetitive isolation of *Aspergillus* spp., often heavy growth, without skin test reactivity, and sometimes positive *Aspergillus* precipitins (11/35 [31%]).[22] In an autopsy series, Young *et al.* found that 8 out of 98 (8%) patients had localized *Aspergillus* bronchitis confirmed histologically and characterized by bronchial casts containing mucus and mycelia.[23] All had superficial erosions and ulcerations and occurred in patients with less neutropenia and less exposure to corticosteroids or antineoplastic agents. Recently Shoseyov *et al.* described six cystic fibrosis (CF) patients with repeat positive sputum cultures of *A. fumigatus* and clinical deterioration, without strong evidence of sensitization to *A. fumigatus*,[28] who responded to antifungal therapy.

We noticed several patients with chronic symptoms who did not fulfill current diagnostic criteria for aspergillosis. Here, we describe the clinical

Figure 1. Drawing of the appearance of the airways from the autopsy of a 3-year-old girl with *Aspergillus* tracheobronchitis who was presumably not immunocompromised.

manifestations, laboratory findings, and treatment outcome of *Aspergillus* bronchitis. We propose criteria for the diagnosis of *Aspergillus* bronchitis.

We recognize that there may be an overlap with invasive *Aspergillus* tracheobronchitis in patients with no preexisting immunocomprise (such as those with severe influenza or given corticosteroids for asthma) and possible confusion with terminology, hence we suggest using bronchitis rather than tracheobronchitis to minimize the potential for confusion.

Methods

Study design

We reviewed the literature for all forms of *Aspergillus* infection of the airway, with a particular emphasis on literature before 1970. A retrospective, observational cohort study was performed. Chart reviews were undertaken of patients who did not fit current diagnostic criteria for allergic, chronic, and invasive aspergillosis identified over seven years from over 400 patients referred to the National Aspergillosis Center until May 2011. Demographic data, comorbidity, clinical presentations, radiological, bronchoscopic, microbiological, and serological findings were collected on patients identified to have *Aspergillus* bronchitis. One patient has previously been reported.[29]

Definitions

Aspergillus bronchitis was defined as (i) symptomatic chronic lower airway disease (symptoms of "chronic bronchitis"[30]), (ii) detection of *Aspergillus* spp. in sputum or BAL by culture or real-time val-

idated PCR, and (iii) detection of IgG antibodies to *Aspergillus* spp. Patients were excluded if they fulfilled the diagnostic criteria for an established fungal-related disease (i.e., chronic pulmonary aspergillosis with or without aspergilloma (CPA), allergic bronchopulmonary aspergillosis (APBA), severe asthma with fungal sensitization (SAFS), invasive aspergillosis (IA)). Some patients with dual or triple *Aspergillus* diseases were excluded. Patients with known CF, neutropenia, or other established profound immune deficit (HIV/AIDS, systemic immunosuppressive agents apart from low-dose steroids to treat respiratory disease exacerbations) were also excluded. Bronchiectasis was defined as bronchial dilatation with respect to the accompanying pulmonary artery (signet ring sign), lack of tapering of bronchi, and identification of bronchi within 1 cm of the pleural surface.[31]

Data collection

Demographic data, underlying diseases, risk factors, and clinical presentations along with radiological, bronchoscopic, microbiological, and serological findings were correlated. Comorbidities were reviewed, with emphasis on immune status. Patients' breathlessness was assessed on the Medical Research Council (MRC) dyspnea scale.[32] Based on this scale, subjects are scored from 1 to 5 on a worsening scale of perceived breathlessness (1 is normal; 5 is severely breathless with trivial activity).

Laboratory methods

Aspergillus precipitin titers were measured using a long-established in-house precipitins IgG assay, as described elsewhere.[33] *Aspergillus* IgG and IgE and total IgE were tested by ImmunoCap® (Phadia, Sweden). Fungal culture of sputum was according to standard UK methods.[34] Sputum was digested with Sputasol (ratio 1:1), then vortexed and 10 μL streaked on two Sabouraud dextrose plates. Plates were incubated for seven days at 30 °C and 37 °C. DNA extraction from sputum was performed from 0.5 mL to 3 mL of sample immediately after the samples were received according to the MycXtra® fungal DNA extraction kit manual (Myconostica, Manchester, UK) using the BBL® Mycoprep™ Specimen Digestion/Decontamination kit (Becton Dickinson, Oxford, UK).[8] Real-time PCR was done with the commercially available MycAssay™ *Aspergillus* (Myconostica) assay.[8] Minimum inhibitory concentrations (MICs) to triazoles were determined by EUCAST methodology as described previously.[35] Antifungal therapeutic drug monitoring was routine for all azole therapies as described elsewhere.[36,37] Serum MBL concentrations were determined by enzyme-linked immunosorbent assay (ELISA) (MBL Oligomer ELISA Kit, BioPorto Diagnostics, DK) with an upper and lower reported detection limit of 4.00 and 0.05 mg/L, respectively.[38]

Response to therapy

Clinical response to antifungal treatment was assessed by course after two months of therapy and at the end of treatment. We excluded patients who received less than four weeks of therapy from response analysis. Duration of treatment along with relapse rate and the need of retreatment were noted. Reasons for the end of therapy, including planned end of therapy, convincing good clinical response or discontinuation due to drug-related adverse events were reviewed.

Results

We identified 17 patients, 14 women, ranging in age from 30 to 76 years (mean 57) who fulfilled our criteria for *Aspergillus* bronchitis. Nearly all patients had concurrent pulmonary or airways disease, most commonly bronchiectasis ($n = 12/14$ (86%)). Six patients were on long-term oral prednisolone, three at ≥ 10 mg daily, one was receiving infliximab, and 12 (70%) were taking inhaled corticosteroids (Table 2). Numerous other comorbidities were seen; two patients had no evidence of underlying disease or were receiving immunosuppressive agents. Mannose binding lectin deficiency was found in 9/16 patients (56%) (Tables 2 and 3).

The clinical presentation was mainly that of persistent cough with sputum production, with frequently recurring exacerbations. Patients initially reported main complaints to be productive cough ($n = 16$, 94%), excessive tenacious sputum ($n = 8$, 47%), severely limiting shortness of breath (MRC dyspnea score ≥ 4) ($n = 8$, 47%), recurrent chest infections ($n = 7$, 41%), extreme fatigue/malaise ($n = 4$, 23%), weight loss more than 3 kg ($n = 3$, 17%), hemoptysis ($n = 1$, 6%), and mucoid impaction requiring urgent bronchoscopy ($n = 1$, 6%).

Bronchoscopy was undertaken in seven patients (Table 4). The bronchoscopic appearances varied widely, from localized areas of inflammation and contact bleeding through marked mucus plugging

Table 2. Underlying pulmonary disease, comorbid conditions, and mannose binding lectin levels in patients with *Aspergillus* bronchitis

Underlying diseases	Number affected (%)
Pulmonary disease	$n = 17$
COPD[a]	6 (35)
Asthma[a]	4 (23)
Bronchiectasis[b]	12/14 (86)
Mucus impaction[b]	2 (12)
Lung cancer	1 (6)
Oral corticosteroids >10 mg/day	3 (18%)
Oral corticosteroids <10 mg/day	3 (18%)
Infliximab	1 (6%)
Inhaled corticosteroids	12 (70%)
Breast cancer radiotherapy	2 (12%)
Hyperthyroidism	2 (12%)
Gamma-IFN production low	1 (6%)
Alpha1 antitrypsin deficiency	1 (6%)
Type II diabetes mellitus	1 (6%)
Hypogammaglobulinemia	1 (6%)
Fibromyalgia	1 (6%)
Irritable bowel syndrome	1 (6%)
No comorbidity	2 (12%)
Mannose binding lectin levels (mg/L)	$N = 16$
>1 (normal)	7 (44%)
>0.5–< 1 (possibly low)	3 (18%)
>0.1–< 0.5 (low)	4 (24%)
<0.1 (undetectable)	2 (12%)

[a]COPD and asthma noted if severe and long established, otherwise probably underestimated.
[b]Bronchiectasis and mucus impaction as determined by HRCT.
COPD, chronic obstructive pulmonary disease; IFN, interferon.

to near normality. Patient 6 had four bronchoscopies to remove tenacious sputum. The histological features of transbronchial biopsy of patient 2 are seen in Figure 2. Chest imaging including CT scanning was noncontributory other than demonstrating bronchiectasis and underlying lung pathology.

At least one, and usually multiple, respiratory cultures were positive in all 17 patients. *A. fumigatus* was grown in 13 patients, *A. niger* in three and *A. terreus* in one. Two patients grew three different species at different times. One isolate was itraconazole and voriconazole resistant, but not all were susceptibility

Table 3. Characteristics of each patient with *Aspergillus* bronchitis

No.	Demographics	Presentation	Comorbidities	Therapies	Bacterial culture	*Aspergillus* culture	*Aspergillus* PCR	IgG/ Precipitins[a]	Itraconazole weeks
1	M/72	Weight loss, productive cough, severe SOB	COPD, right lung carcinoma, right pneumonectomy	Inhaled steroids, systemic steroids	*S. liquefi-cans, C. parapsilo-sis*	*A. fumigatus* BAL	Positive	IgG 25, precipitins 1/2	12
2	M/30	Cough, tenacious phlegm, recurring chest infections, extreme fatigue	Asthma	Inhaled steroids	*H. influenzae, P. aeruginosa*	*A. fumigatus* BAL + sputum	Positive	IgG 90	52
3	F/30	Low grade fever, cough, excessive sputum, malaise	Asthma, frequent chest infections	Inhaled steroids, systemic steroids	Negative	Sputum – repeatedly *A. niger*	ND	IgG 19	1
4	F/62	SOB, cough, yellow, thick sputum, 8–9 chest infections/year, weight loss 15 kg/year	Previous breast cancer	Inhaled steroids	*P. aeruginosa, S. pneumoniae*	*A. fumigatus* sputum	Positive	IgG 4	9
5	F/58	SOB, exercise tolerance 30 yards, thick green sputum	COPD, fibromyalgia	Inhaled steroids	*S. aureus, H. influenzae*	*A. fumigatus, A. niger*, and *A. terreus*	Positive	IgG 82	20
6	F/58	Recurrent chest infections, thick phlegm, occasional night sweats, persistent cough	Hyperthyroidism, asthma	Inhaled steroids	Negative	*A. fumigatus* sputum + BAL	Sputum posit, BAL neg	IgG 82	26
7	F/58	SOB, MRC 4, frequent chest infections, massive sputum production	Irritable bowel syndrome	No steroids noted	*P. aeruginosa*	*A. fumigatus*	Positive	IgG 72, precipitins 1/1	52
8	F/46	Recurrent chest infections, severe SOB MRC 4–5, no sputum, no cough, no haemoptysis, weight loss 15 kg	Previous breast cancer with radiotherapy, COPD	Periodic systemic steroids for chest symptoms	Negative	*A. fumigatus* BAL	Positive sputum + BAL	IgG 144	0
9	M/58	Recurrent chest infections and hospital admissions, MRC 5, home oxygen therapy	Severe COPD, type 2 diabetes, ischemic heart disease	Systemic steroids long term, inhaled steroids	*P. aeruginosa*	Negative	Positive sputum	IgG 65	3
10	F/68	Recurrent chest infections, severe cough, voluminous sputum, SOB MRC 5, exercise tolerance 20 yards, weight loss	Severe COPD, osteoporosis, vertebral fractures	Short systemic steroids, inhaled steroids	*P. aeruginosa* + coliforms	*A. fumigatus*	Unresolved – inhibition	IgG 100	20

Continued

Table 3. *Continued*

No.	Demographics	Presentation	Comorbidities	Therapies	Bacterial culture	*Aspergillus* culture	*Aspergillus* PCR	IgG/ Precipitins[a]	Itraconazole weeks
11	F/73	Recurrent chest infections, voluminous green sputum, SOB MRC 3, wheeze	Previous cavitating pulmonary tuberculosis, asthma, hypertension, hypercholes- terolemia	Inhaled steroids	*P. aerugi- nosa, H. influenzae*	*A. fumigatus*	ND	Precipitins 1/4	52
12	F/42	Frequent chest infections, recurring hemoptysis	Bronchiectasis, alpha1 antitrypsin deficiency	Probably no steroids	*S. aureus, H. influenzae*	*A. fumigatus*	Positive	IgG 47, pre- cipitins 1/4	4
13	F/59	Low-grade fever, frequent chest infections, green sputum, normal exercise tolerance	Asthma, MAI infection, latent TB	Inhaled steroids	MAI, *S. aureus, P. aerugi- nosa, Candida*	*A. fumigatus*	Negative	IgG 57	26
14	F/73	Bronchial mucoid impaction, lobe collapse, tenacious thick sputum	Nil	Inhaled steroids	Negative	*A. fumigatus* BAL	ND	IgG 137, precip- itins 1/1	0
15	F/76	SOB, cough, sputum hyperproduction	Rheumatoid arthritis	Infliximab	*H. influenzae*	Negative	Positive	IgG 105	0
16	F/48	SOB, cough, fatigue, low-grade fever, fatigue, sticky mucous sputum	Gamma- interferon deficiency, adrenal insufficiency	Inhaled steroids, cortisole replacement	Negative	BAL *A. fu- migatus,* sputum *A. fumi- gatus,* and *A. niger*	Positive	*A. niger* IgG positive, *A. fumigatus* IgG 16, negative precip- itins	38
17	F/67	SOB, exercise tolerance 30 yards, white sputum	COPD	Hyperthyroidism, hypogamma- globuli- naemia, steroid inhalers + short course of oral steroids	*P. aerugi- nosa, S. mal- tophilia*	*A. fumigatus*	Positive	IgG 86, pre- cipitins 1/2	2

[a]The figures after precipitins refers to the dilution titre that remains positive. COPD, chronic obstructive pulmonary diseases; SOB, shortness of breath; MRC refers to the Medical Research Council dysnea score (range 1 to 5; 1 = normal, 5 = breathless eating, talking, etc.) MAI, *Mycobacterium avium intracellulare*; ND, not done.

tested. Real-time PCR was positive in 12/14 (86%) samples taken in Manchester.

Elevated *A. fumigatus*-specific IgG antibody (Im-munoCap) was detected in 12 (range 47–144 mg/L (mean 88.9 mg/L)) and precipitins in six patients (range 1:1–1:4 titers), in three cases without posi-tive ImmunoCap antibodies. Three patients had de-tectable *A. niger* IgG (done in three different external laboratories) and one had detectable *A. terreus* IgG antibody. In two of those with positive *A. niger* IgG antibodies, *A. niger* complex was the most frequent isolate from respiratory samples. Total serum IgE was ≤ 100 in 11 (65%) patients and ranged up to 760 KIU/L. This last patient did not fulfill criteria for ABPA and had a rapid and dramatic response to itraconazole. *Aspergillus*-specific IgE was ≤ 0.4

Table 4. Findings in the seven patients who underwent bronchoscopy

Patient	Bronchoscopic findings	Biopsy	Mycology and bacteriology
1	Inflamed left main bronchus, with several areas of contact bleeding; distal airways normal	Hyphae invading bronchial mucosa	BAL *A. fumigatus, Candida parapsilosis*, and *Serratia liquefcans* cultured; *Aspergillus* PCR sputum positive
2	Cream-colored mucosal irregularity LUL, sticky pale secretions throughout, mucosa friable and easy bleeding, neutrophil inflammation, plaque-like lesions with ulcerations due to *Aspergillus*	Focal squamous metaplasia. Area of ulceration covered by granulation tissue, with a superficial layer of fungal hyphae. Chronic inflammatory cell infiltrate without eosinophilia.	BAL *A. fumigatus*. Also grew *H. influenzae* and *P. aeruginosa* from sputum
6	Edematous narrowed airways—suggestive of bronchitis/asthma	ND	BAL *A. fumigatus*; *Aspergillus* PCR negative. Sputum culture *A. fumigatus*, *Aspergillus* PCR sputum positive
8	Normal bronchoscopy	ND	BAL *A. fumigatus*; *Aspergillus* PCR positive. Sputum *Aspergillus* PCR sputum positive
12	Normal apart from tracheobronchopathy chondroplasitica, Bronchoscopy having stopped posaconazole.	ND	Sputum culture *A. fumigatus*, *Aspergillus* PCR sputum positive. Azole resistant. Also grew *S. aureus* and *H. influenzae*
14	Thick tenacious sputum, plugging the LLL bronchus	ND	BAL fungal culture negative; *Aspergillus* PCR positive
16	Normal with three tiny black spots visible on subsequent bronchoscopy (on treatment)	ND	BAL *A. niger* cultured and *Aspergillus* PCR positive

KIU/L in 12 (71%) and 8.7 KIU/L was the highest value.

Patients ($n = 14$) were started on oral itraconazole, 200 mg twice a day or voriconazole ($n = 1$), and dose was adjusted in five patients according to drug levels. The intended duration of the first course of antifungal therapy was arbitrarily predetermined to be four months. For relapse, longer courses for up to 52 weeks were given, or longer in the case of further relapses. In the case of itraconazole intolerance ($n = 7$, 50%), voriconazole was substituted in three patients but had to be discontinued due to side effects in all three after 16, 6, and 36 weeks. Posaconazole was started in two patients, one with itraconazole and voriconazole resistance, the other with significant intolerance to both agents. Both

patients responded to posaconazole, but discontinuation in one patient after four months resulted in relapse, which responded to reinstitution of long-term posaconazole.[29]

Antifungal response as initial treatment was assessable in 12 patients (11 with itraconazole). After two months of antifungal therapy, major symptomatic improvement was reported by six patients (50%), mild-to-moderate improvement was reported by five (42%) patients, and no improvement was reported in one who had itraconazole resistance. Seven patients maintained the improvement gained during therapy after discontinuing itraconazole and without relapse after 9–52 weeks of therapy (median 26 weeks). Five patients relapsed, two after itraconazole, two

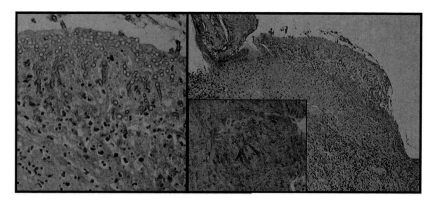

Figure 2. Transbronchial biopsy of patient 2 stained with H&E showing focal squamous metaplasia in the superficial layers of the bronchial epithelium (left panel). An area of ulceration covered by granulation tissue, with a superficial layer of fungal hyphae is seen from 12–3:00 o'clock. Deeper in the biopsy there is a chronic inflammatory cell infiltrate without eosinophilia. The inset shows septate branching hyphae typical of Aspergillus superficially within bronchial tissue.

after voriconazole, and one after posaconazole and then reresponded (major response) after restarting antifungal therapy. Three patients had a full recovery, one having relapsed after itraconazole and responded to voriconazole. At the time of analysis six patients continue on antifungal therapy with good control of their symptoms.

There were seven patients whose main complaint upon presentation included frequent lower respiratory tract infections, poor response to antibiotics, and rapid relapse after antibiotic discontinuation. Four of these patients showed major response to antifungal therapy with resolution of respiratory symptoms, and partial response was seen in two cases; one did not tolerate antifungal therapy. Two of those with major responses relapsed after discontinuing antifungal therapy with rapid resolution of symptoms after restarting antifungals. Of these seven patients, five were chronically colonized with *Pseudomonas aeruginosa,* one with *Staphylococcus aureus,* and one had no evidence of bacterial colonization of the bronchial tree. Six of the patients had bronchiectasis.

Discussion

There were two common presentations of *Aspergillus* bronchitis: recurrent chest infections that were treated with repeated unsuccessful courses of antibiotic treatment, and significant breathlessness with mucoid impaction. In occasional instances in the literature, symptoms followed a significant presumptive exposure to airborne fungi, but not in our patients. In several of these patients, antifungal therapy was given as a therapeutic trial after no response to multiple antibacterial regimens and their clinical response was surprisingly prompt, some with subsequent relapse. Before itraconazole availability, some responses were noted to oral iodide therapy.[12,18]

A key finding in all the cases was repetitive identification of *Aspergillus* spp. in respiratory samples by culture or real-time PCR. In order to infer a persistent disease process of the airways, repeated detection of that pathogen from the airways is essential. In the early literature, microscopy alone or combined with culture was sometimes used for diagnosis. The significance of a single isolation of *Aspergillus* from sputum can be difficult to assess, as it may be normal flora for some people[8,39] and/or represent sample or agar plate contamination. Real-time PCR is much more sensitive than culture and is quantitative.[8,40–42] High signals were characteristic of these patients, consistent with heavy *Aspergillus* loads in the airways. The line between long-term colonization and airway infection in the context of structural airway damage and immunological deficit is often difficult to draw and may not be static, with excursions to a more invasive or superficially invasive form at periods of increasing immunosuppression. Therein lies the clinical challenge: to distinguish *Aspergillus* colonization from airway infection in patients with persistent symptoms.

Evidence in favor of *Aspergillus* bronchitis as a discrete clinical entity includes finding overt disease on bronchoscopy (Fig. 1). Two of our patients had bronchial biopsy that showed localized invasion of hyphae. Both patients had months of symptoms, which is not consistent with the clinical course of invasive *Aspergillus* tracheobronchitis. In pathological specimens, Symmers described mild bronchial inflammation and/or excess production of mucus without invasion,[21] whereas Young *et al.*[23] found bronchial casts containing mucus and mycelia with superficial mucosal erosion and ulceration. Bronchial erythema and excess mucus is also well recognized in orthoptic lung transplant recipients[6] and may precede invasive *Aspergillus* tracheobronchitis[7] or ulcerative *Aspergillus* tracheobronchitis.

Other potentially supportive data include a humoral antibody response. *Aspergillus* precipitating IgG antibodies were detected in six patients, while specific IgG were raised in 12 patients out of 14 with *A. fumigatus* bronchitis. Only four patients had both raised *Aspergillus*-specific IgG and positive precipitins. Three of our patients were infected with non-*fumigatus* Aspergilli (*A. niger* and *A. terreus*), which may not reliably induce cross-reacting antibodies and would require alternative (and less well studied) IgG testing. These antibody tests were introduced in the 1960s and so cases prior to this were not tested.

On the basis of our observations, we propose criteria for the diagnosis of *Aspergillus* bronchitis in patients without significant immunocompromise

Table 5. Proposed criteria for the diagnosis of *Aspergillus* bronchitis

	Aspergillus colonization	*Aspergillus* bronchitis
Essential criteria		
1. Microbiology	Single sputum or PCR positive	Repeat sputum culture or PCR positive for *Aspergillus* sp.
2. Symptoms	Lack of new substantial symptoms	Chronic (> 4 weeks) pulmonary symptoms Possibly systemic symptoms
3. Other forms of aspergillosis	Patient does not fulfill the criteria for invasive, allergic, or chronic aspergillosis (appropriate tests done)	
4. Immune system deficiency[a]	No overt immunocompromise, such as recent chemotherapy, transplantation or AIDS, where a more severe and invasive form of *Aspergillus* tracheobronchitis usually develops	
Supportive criteria		
5. Serology[b] (IgG or precipitins)	*Aspergillus* antibody negative in serum	*Aspergillus* IgG antibody detectable in serum
6. Bronchoscopy findings	Normal or fixed structural abnormality	Mucoid impaction, thick tenacious sputum with bronchial plugging, bronchial erythema (touch bleeding) and/or ulceration Superficial invasion of mucosa by *Aspergillus* hyphae
7. Response to therapy	Equivocal or no response in given	Good response to an eight-week course of antifungal therapy

[a]If the patient otherwise fulfills the criteria for *Aspergillus* bronchitis but is significantly immunocompromised, even temporarily such as receiving high-dose corticosteroids for an asthma exacerbation, the term invasive *Aspergillus* tracheobronchitis should be applied.
[b]This could be regarded as an essential criterion, if superficial mucosal invasion is not demonstrated, but the lack of study of *Aspergillus* IgG antibody testing without a cutoff prevents this from being adopted in the current state of knowledge.

(Table 5). Persistent respiratory symptoms unresponsive to antibiotics in either nonimmuno-compromised patients or patients with minor immunocompromising factors with *Aspergillus* spp. detectable in sputum (culture or real-time PCR) could indicate *Aspergillus* bronchitis. Bronchoscopy findings of mucoid impaction and/or localized ulceration with superficial invasion by *Aspergillus* hyphae are characteristic. Detectable *Aspergillus* IgG antibodies is a supportive criterion. Further studies are needed to define a cut-off for *Aspergillus* IgG antibody. Coinfection with bacterial pathogens appears to be common in *Aspergillus* bronchitis.

The precise nature of the local immune deficit in *Aspergillus* bronchitis requires better definition. Variable presentations of human encounters with pathogens that are abundantly present in the environment is to be expected. *Aspergillus* bronchitis occurs primarily in patients with bronchiectasis who are not overtly immunocompromised. Yet, some impairment of immune defense is likely. After inhalation, alveolar macrophages eliminate *Aspergillus* spores by phagocytosis.[43] Epithelial impairment in emphysema and the use of corticosteroids, especially in the inhaled form, have been suggested as instrumental in the increase of IA in the immunocompetent population.[44] Guinea noted systemic corticosteroid use to be an independent risk factor for invasive aspergillosis in COPD in critical care settings.[9] The TORCH study noted an increase in pneumonia related to high-dose inhaled corticosteroids, but not opportunistic lung infection.[45] The Th1/Th2 balance and IL-17/IL-23 immunotolerance regulation appear also to be pivotal in the defense against *Aspergillus*,[46] with mucosal immunity particularly dependent on Th17 responses. We found over 50% of our patients to have MBL deficiency. MBL could be a disease modifier in the context of *Aspergillus* bronchitis, as has been suggested for CPA and/or allergic aspergillosis in cystic fibrosis.[38,47] *A. fumigatus* also appears to have potent capacity to induce hyperproduction of mucus in the bronchial epithelium via the activation of the TACE/TGF-α/epidermal growth factor receptor pathway.[48] Biofilm formation could also be relevant to pathogenesis.

Thus, we have described a group of patients with underlying structural and immunological compromise of varying severity, who present with difficult-to-treat and readily relapsing symptoms of chronic bronchitis combined with microbiological and immunological evidence of *Aspergillus* infection. While not immunocompromised in the sense of common clinical parlance, these patients usually have one or more minor immune deficits. Most of these patients benefit from a course of oral antifungal therapy, though many relapse. Additional work is necessary to optimize therapy, establish an appropriate duration, and define groups of patients who need long-term therapy. *Aspergillus* bronchitis should be regarded as a chronic superficial airway infection similar to bacterial bronchitis. Introducing this category of disease enables a distinction between asymptomatic fungal colonization and allergic bronchopulmonary aspergillosis, facilitating treatment of those afflicted. Further research into the category of subtle underlying defects of immune response and patterns of interactions with opportunistic pathogens is needed.

Acknowledgments

The primary data collection and analysis was undertaken by Drs. A.C. and D.W.D. Dr. S.M. reviewed the literature exhaustively. Drs. R.J.B.-T., T.F., and C.G.B. contributed to the patient's care and provided valuable clinical insights. The first draft was written by Dr. A.C., all authors contributed to the final draft. Dr. D.W.D. is the guarantor of the paper's veracity. We are indebted to our colleagues for referring patients, to Mrs. Chris Harris for collating notes and results, to Dr. Paul Bishop for photographing the transbronchial biopsy of patient 2 and the local primary Care Trusts for funding voriconazole and posaconazole for these patients. Drs. A.C., C.G.B., and T.F. were funded by the National Commissioning Group and the Medical Research Council. Dr. S.M. received a travel scholarship from the Malaysian Government to visit the National Aspergillosis Center.

Conflicts of interest

The authors declare no conflicts of interest.

References

1. Hope, W.W., T.J. Walsh & D.W. Denning. 2005. The invasive and saprophytic syndromes due to *Aspergillus* spp. *Med. Mycol.* **43:** 207–238.
2. Segal, B.H. 2009. Aspergillosis. *N. Engl. J. Med.* **360:** 1870–1884.

3. Fedorova, N.D., N. Khaldi, V.S. Joardar, *et al.* 2008. Genomic islands in the pathogenic filamentous fungus *Aspergillus fumigatus*. *PLoS Genet.* **4:** e1000046.

4. Denning, D.W. 1995. Commentary: unusual manifestations of aspergillosis. *Thorax* **50:** 812–813.

5. Kramer, M.R., D.W. Denning, S.E. Marshall, *et al.* 1991. Ulcerative tracheobronchitis after lung transplantation. A new form of invasive aspergillosis. *Am. Rev. Respir. Dis.* **144:** 552–556.

6. Mehrad, B., G. Paciocco, F.J. Martinez, *et al.* 2001. Spectrum of *Aspergillus* infection in lung transplant recipients: case series and review of the literature. *Chest* **119:** 169–175.

7. Wu, N., Y. Huang, Q. Li, *et al.* 2010. Isolated invasive Aspergillus tracheobronchitis: a clinical study of 19 cases. *Clin. Microbiol. Infect.* **16:** 689–695.

8. Denning, D.W., S. Park, C. Lass-Florl, *et al.* 2011. High frequency triazole resistance found in non-culturable *Aspergillus fumigatus* from lungs of patients with chronic fungal disease. *Clin. Infect. Dis.* **52:** 1123–1129.

9. Guinea, J., M. Torres-Narbona, P. Gijón, *et al.* 2010. Pulmonary aspergillosis in patients with chronic obstructive pulmonary disease: incidence, risk factors, and outcome. *Clin. Microbiol. Infect.* **16:** 870–877.

10. Fairs, A., J. Agbetile, B. Hargadon, *et al.* 2010. IgE sensitization to *Aspergillus fumigatus* is associated with reduced lung function in asthma. *Am. J. Respir. Crit. Care Med.* **182:** 1362–1368.

11. Wheaton, S.W. 1890. Case primarily of tubercle in which a fungus (aspergillus) grew in the bronchi and lung, simulating actinomycosis. *Trans. Path. Soc.* **41:** 34–37.

12. Hoxie, G.H. & F.C. Lamar. 1912. Fungous tracheobronchitis. *JAMA* **85:** 95–96.

13. Lapham, M.E. 1926. Aspergillosis of the lungs and its association with tuberculosis. *JAMA* **87:** 1031–1033.

14. Wahl, E.F. 1928. Primary pulmonary aspergillosis. *JAMA* **91:** 200–202.

15. Castellani, A. 1928. Fungi and fungous diseases. *Arch. Dermatol. Syphilis.* **17:** 61.

16. Schneider, L.V. 1930. Primary aspergillosis of the lungs. *Am. Rev. Tuberc.* **22:** 267–270.

17. Fawcitt, R. 1936. Fungoid conditions of the lungs—Part I. *Br. J. Radiol.* **9:** 172–195.

18. Von Orstrand, H.S. 1940. Pulmonary aspergillosis. *Cleveland Clin. Q.* **7:** 66–73.

19. Hinson, K.F.W., A.J. Moon & N.S. Plummer. 1952. Bronchopulmonary aspergillosis. A review and a report of eight new cases. *Thorax* **7:** 317–333.

20. Riddell, R.W. & Y.M. Clayton. 1958. Pulmonary mycoses occurring in Britain. *Br. J. Tuberc. Dis. Chest* **52:** 34–44.

21. Symmers, W.S. 1962. Histopathologic aspects of the pathogenesis of some opportunistic fungal infections, as exemplified in the pathology of aspergillosis and the phycomycetoses. *Lab. Invest.* **11:** 1073–1090.

22. Campbell, M.J. & Y.M. Clayton. 1964. Bronchopulmonary aspergillosis. A correlation of the clinical and laboratory findings in 272 patients investigated for bronchopulmonary aspergillosis. *Am. Rev. Respir. Dis.* **89:** 186–196.

23. Young, R.C., J.E. Bennett, C.L. Vogel, *et al.* 1970. Aspergillosis. The spectrum of the disease in 98 patients. *Medicine* **49:** 147–173.

24. Berlinger, N.T. & T.J. Freeman. 1989. Acute airway obstruction due to necrotizing tracheobronchial aspergillosis in immunocompromised patients: a new clinical entity. *Ann. Otol. Rhinol. Laryngol.* **98:** 718–720.

25. Denning, D.W., S. Follansbee, M. Scolaro, *et al.* 1991. Pulmonary aspergillosis in AIDS. *N. Engl. J. Med.* **324:** 654–662.

26. Kemper, C.A., J.S. Hostetler, S.E. Follansbee, *et al.* 1993. Ulcerative and plaque-like tracheobronchitis due to infection with Aspergillus in patients with AIDS. *Clin. Infect. Dis. Sep.* **17:** 344–352.

27. Sambatakou, H. & D.W. Denning. 2005. Invasive pulmonary aspergillosis transformed by immune reconstitution in AIDS into fatal mucous impaction. *Eur. J. Clin. Microbiol. Infect. Dis.* **24:** 628–633.

28. Shoseyov, D., K.G. Brownlee, S.P. Conway & E. Kerem. 2006. *Aspergillus* bronchitis in cystic fibrosis. *Chest* **130:** 222–226.

29. Howard, S.J., A.C. Pasqualotto & D.W. Denning. 2010. Azole resistance in ABPA and *Aspergillus* bronchitis. *Clin. Microbiol. Infect.* **16:** 683–688.

30. Global Strategy for the Diagnosis, Management, and Prevention of Chronic Obstructive Pulmonary Disease, Updated 2010, © 2010 Global Initiative for Chronic Obstructive Lung Disease, Inc.

31. Hansell, D.M., A.A. Bamkier, H. MacMahon, *et al.* 2008. Remy J. Fleischner Society: glossary of terms for thoracic imaging. *Radiology* **246:** 697–722.

32. Bestall, J.C., E.A. Paul, R. Garrod, *et al.* 1999. Usefulness of the Medical Research Council (MRC) dyspnoea scale as a measure of disability in patients with chronic obstructive pulmonary disease. *Thorax* **54:** 581–586.

33. Jain, L.R. & D.W. Denning. 2006. The efficacy and tolerability of voriconazole in the treatment of chronic cavitary pulmonary aspergillosis. *J. Infect.* **52:** e133–e137.

34. Standards Unit, Evaluations and Standards Laboratory, Centre for Infections. Investigation of bronchoalveolar lavage, sputum and associated specimens; BSOP57. Available at: www.hpa-standardmethods.org.uk/documents/bsop/pdf/bsop57.pdf. Accessed September 4, 2010.

35. Howard, S.J., D. Cerar, M.J. Anderson, *et al.* 2009. Frequency and evolution of azole resistance in *Aspergillus fumigatus* associated with treatment failure. *Emerg. Infect. Dis.* **15:** 1068–1076.

36. Andes, D., A. Pascual & O. Marchetti. 2009. Antifungal therapeutic drug monitoring: established and emerging indications. *Antimicrob. Agents Chemother.* **53:** 24–34.

37. Hope, WW, E.M. Billaud, J., Lestner & D.W. Denning. 2008. Therapeutic drug monitoring for triazoles. *Curr. Opin. Infect. Dis.* **21:** 580–586.

38. Harrison, E., A. Singh, N. Smith, *et al.* 2012. Mannose binding lectin genotype and serum levels in patients with chronic or allergic pulmonary aspergillosis. *Int. J. Immunogenetics* **39:** 224–232.

39. Lass-Flörl, C., G.M. Salzer, T. Schmid, *et al.* 1999. Pulmonary *Aspergillus* colonization in humans and its impact on management of critically ill patients. *Br. J. Haematol.* **104:** 745–747.

40. Lass-Flörl, C., S.A. Follett, A. Moody & D.W. Denning. 2011. Detection of Aspergillus in lung and other tissue samples using the MycAssay™ Aspergillus real-time PCR kit. *Can. J. Microbiol.* **57:** 765–768.

41. Baxter, C.G., A. Jones, A.K. Webb & D.W. Denning. 2011. Homogenisation of cystic fibrosis sputum by sonication—an essential step for *Aspergillus* PCR. *J. Microbiol. Method* **85:** 75–81.

42. Torelli, R., M. Sanguinetti, A. Moody, *et al.* 2011. Diagnosis of invasive aspergillosis by a commercial real-time PCR assay for *Aspergillus* DNA in bronchoalveolar lavage fluid samples from high-risk patients compared to a galactomannan enzyme immunoassay. *J. Clin. Microbiol.* **49:** 4273–4278.

43. Schaffner, A., H. Douglas & A. Braude. 1982. Selective protection against conidia by mononuclear and against mycelia by polymorphonuclear phagocytes in resistance to *Aspergillus*. Observations on these two lines of defense *in vivo* and *in vitro* with human and mouse phagocytes. *J. Clin. Invest.* **69:** 617–631.

44. Ader, F. 2010. Invasive pulmonary aspergillosis in patients with chronic obstructive pulmonary disease: an emerging fungal disease. *Curr. Infect. Dis. Rep.* **12:** 409–416.

45. Crim, C., P.M. Calverley, J.A. Anderson, *et al.* 2009. Pneumonia risk in COPD patients receiving inhaled corticosteroids alone or in combination: TORCH study results. *Eur. Respir. J.* **34:** 641–647.

46. Brown, G.D., D.W. Denning, N.A.R. Gow, *et al.* Human fungal infections: the hidden killers. *Sci. Transl. Med.* In press.

47. Muhlebach, M.S., S.L. MacDonald, B. Button, *et al.* 2006. Association between mannan-binding lectin and impaired lung function in cystic fibrosis may be age-dependent. *Clin. Exp. Immunol.* **145:** 302–307.

48. Oguma, T., K. Asano, K. Tomomatsu, *et al.* 2011. Induction of mucin and MUC5AC expression by the protease activity of *Aspergillus fumigatus* in airway epithelial cells. *J. Immunol.* **187:** 999–1005.